Easy Things to Make With
CONCRETE & MASONRY

by Richard Day

arco
New York

ACKNOWLEDGMENTS

Thanks is expressed to the following people and organizations for help with this book: Brick and Tile Service, Inc.; Cement and Concrete Association; Robert Cleveland; Jack Cofran; Concrete Machinery Co.; Frank D. Davis Co.; Stuart Day; Delawie, Macy Associates; Hazard Products, Inc.; Tom Miller; Portland Cement Association; Carl Roth; Sakrete, Inc.; Structural Clay Products Inst.; and Jerry Woods.

Published by ARCO PUBLISHING COMPANY, Inc.
219 Park Avenue South, New York, N.Y. 10003

First ARCO Edition, 1972

Copyright © 1971 by Fawcett Publications, Inc.

Library of Congress Catalog Card Number 72-85574
ISBN 0-668-02698-7

Printed in the United States of America

CONTENTS

To make it easy on yourself, you need lots of help with a big project, such as Carl Roth's garage floor-foundation. Ready mix is best for large projects. Truck can dump right in form.

USING CONCRETE
THE EASY WAY

Taking advantage of what machines can do for you makes concrete go best

EASY, IN WORKING with concrete, has got to be relative. Nothing's as easy as lying back and watching a concrete contractor build you a patio. By *easy,* we mean *easier than* and *the easy way.* If you want the quality approach to concrete and masonry, get Fawcett Book 677, THE NEW CONCRETE AND MASONRY GUIDE.

Okay, that understood, how do you go about making and using concrete the easy way? The secret is to take full advantage of machine labor. Here are the ways: use ready-packaged mixes; buy or rent a mixer; use ready mixed concrete; use haul-it-yourself concrete.

Which method you use should depend largely on the job to be done. Ready-packaged mixes are a great convenience for doing small jobs. Anything of less than about two yards of concrete is a likely candidate for the premixes.

For any project of more than a few cubic feet of concrete, you should use a mixer to combine the ingredients. Hand-mixing is hard work. Various sizes of mixers are available. If you don't want to buy, you can easily rent.

Intermediate-sized projects of, say, a half cubic yard to 2 cubic yards are most easily handled by haul-it-yourself concrete, if that's available in your area. A small trailer mixer is filled with concrete mix and you tow it behind your car.

Ready mixed concrete, which is delivered in a big truck-mixer, doesn't become practical until you need at least a cubic yard of mix. When more than 2 cubic yards of concrete are needed at a time, ready mix is the only practical way to go.

Sand-mix is used for topping and casting thin-section projects. You can mix small amounts on the sidewalk with a shovel. Larger amounts should be mixed in a mixer.

The amount of water you add to a mix is important. Follow directions on the sack. Good concrete should stand up in piles like this. You really have to work it into place.

READY-PACKAGED MIXES

Prepackaged concrete mixes can produce excellent concrete. They're available in two kinds: gravel-mix, also called *concrete mix*: and sand-mix, also called *topping*. Gravel-mix is a normal concrete mix containing cement, concrete, sand and gravel, but no water. Sand-mix contains cement and concrete sand, but no gravel.

Gravel-mix is most economical to use, because the stones fill out the mix without much cost. Sand-mix has no gravel filler, and so more of the costly cement is needed per cubic foot of mix than with gravel-mix. Use gravel-mix for any project that has a cross-section great enough to accommodate the gravel. Normally, this would be about 4 inches minimum cross-section. The rule-of-thumb is that the minimum cross-section of your project should be 5 times the size of the largest gravel chunks. Usually in ready-packaged mixes these are ¾ inch. That's why the 4-inch cross section requirement. But gravel-mix may be used successfully down to about half that thickness.

Sand-mix should be used when the cross-section is too small for gravel-mix. This is the case with most stepping stones, topping mixes and many precast objects.

Prepackaged mixes are available in various-sized bags, usually 25-, 45-, 60- and 90-pound. A 60-pound bag of premix gives about ½ cubic foot of concrete. With ready-packaged mixes, the job of selecting and proportioning materials has already been done for you. All you do is add water and mix. Directions on the bag even tell you how much water to add.

QUALITY OF PREPACKS

One problem in using ready-packaged mixes is the tremendous variation in quality from brand to brand. There are no standardized requirements for quality. Buy only a well known brand. The best-known one is *Sakrete*.

When you use a premix, you can judge its quality reasonably well. The good ones contain enough cement to make lots of cement paste that will surround the particles of sand and gravel. The others skimp on cement so that the mix appears sandy or gravelly. It will be hard to place, tough to finish and weak.

YOUR RESPONSIBILITY

When you use ready-packaged mixes, be sure to mix the contents of each bag completely before you make concrete with any portion of it. The reason is that, although the ingredients are mixed throughly before packaging, they usually settle out in shipment and storage. Thus most of the sand and cement can be in one end of the bag and most of the gravel can be in the other end. If you dip out a shovel full of mix without mixing the whole bag, you can get a heck of a variation in proportions from top to bottom of the bag.

Another quality measure that you, not the manufacturer, control is the amount of water added. In any concrete you make, the less water you use, the stronger and more durable the concrete will be. If you want really strong concrete, use only enough water to dampen the mix. You'll have to tamp it into place, and you won't get much of a finish on it, but it will be strong concrete.

The quantities of mixing water given on the bag give a combination of good working plus strength and durability. Stick by them to the point of fussiness and you'll be way ahead of the game. Mark off a container with a piece of plastic tape and use it for measuring a one-bag batch of water.

If strength and durability aren't as important to you as ease of placing, you can add a little more water to the mix. Just realize that you are sacrificing really good concrete to get soupy, easy-to-form concrete that almost flows into place.

HIGH COST

Probably the biggest drawback to using ready-packaged mixes is their high unit cost, meaning cost per cubic yard of concrete. Figured out, it comes to something like $40 a yard. You can buy ready mix, which comes delivered to you already mixed and dumped in place for something like $25 a cubic yard in large quantities. However, if you only need one cubic yard, your ready mix dealer may add on a charge for delivery. Many do on small amounts. Then the cost will come to about the same.

MACHINE-MIXING

You have two choices of mixing methods for ready-packaged concrete: mix by hand and machine-mix. All but the smallest projects call for machine-mixing. It saves your time and energy for placing and finishing the concrete. It also does a better job of combining ingredients. With a decent-sized concrete mixer—rented or owned—you can probably handle the mixing, placing and finishing of one to two cubic yards of concrete a day.

If you will have much use for it, by all means get your own mixer. They don't cost a lot. Consider it a lifetime investment. Renting it to others can help you recoup on your investment, even make a profit.

Mixer sizes vary from handy 5-gallon-pail ones to large 7-cubic-foot contractor types. For $100 you can get a moderate-sized one with electric motor drive that will serve well.

Mixer size is always given as drum capacity. Look on the machine's nameplate. Actual concrete mixing capacity is 60 percent of drum size. A 5-cubic-foot mixer will mix about 3 cubic feet of concrete, which needs half a sack of cement. This is otherwise called a *half-bag* mixer. It's a good size, but big.

If you don't use ready-packaged mixes you'll have to buy portland cement, concrete sand and ¾-inch-maximum size graded gravel or crushed stones and proportion your own concrete. A widely used rule-of-thumb for batching is 1:2:3. This means one part cement, two parts sand and three parts gravel. If you're careful to get equal-sized shovelfuls, you can count as you shovel the ingredients into the mixer. Just enough water should be added to make a workable mix that you can place.

LOADING THE DRUM

In what order you put the materials in the drum helps the drum to stay clean during use. Do it this way while the mixer runs: (1) pour in about one-tenth of the water; (2) add the gravel; (3) shovel in the sand; (4) put in the cement; (5) finally add the rest of the water.

Concrete should mix at least a minute after all the ingredients are in the drum. Three minutes of mixing is better.

To clean the drum after use, shovel in some stones and add lots of water, hosing down the inside and outside of the drum. Never leave mortar deposits on and around the mixing drum.

HAND-MIXING

If you hand-mix your concrete, it goes much easier in a wheelbarrow or other large enclosed container. Lacking that, you can hand-mix on a flat slab, such as a sidewalk.

First place the materials in a pile and mix them dry until they look uniform. Use a flat shovel or hoe. Make a depression in the center of the pile and add water a little at a time, mixing the water in. Mix thoroughly after the last water is in, then your concrete is ready to use.

HAUL-IT-YOURSELF

If you can find a place nearby to buy

Portland Cement Assn.

To machine-mix your own concrete, buy portland cement, sand and gravel and batch them into the mixer. One way to handle the cement is to open the sacks in a wheelbarrow. Another way to handle cement-batching of a half-bag batch is by cutting the bag across the center with a shovel and lifting it. Add one half to mixer (top right). Shovel sand from the pile into the mixer, counting the number of shovels. Other materials are batched by shovel-count too by the 1:2:3, cement:sand:gravel, formula. Water should be measured out in a marked can. When the sand is of average wetness, no more than 11 qt. of water should be added to a mix containing a half-bag of cement. If the materials are proportioned correctly, the mix should look like this (bottom, left) in the mixer drum. A gravelly or sandy appearance is corrected by varying the amount of gravel added. Too wet, this mix (bottom right) flows out into a puddle when placed. Note how the stones separate from the paste.

7

After mixing for three minutes, dump batch out and wheel it where it's needed. A rubber-tired wheelbarrow gives the mix a smoother ride to prevent settling. Putting plywood down for the ready mix truck to drive on can save cracks in a driveway. Roth, a professional engineer, designed his driveway for light loads. Plywood spreads the load. Concrete tools are used in placing and finishing. Wheelbarrow and flat shovel are used here by Jack Cofran to place concrete in the forms for a reinforced well pit cover. Using his shovel, Cofran consolidates concrete against the edge forms and around the rebars. Good concrete will not flow into place, must be compacted. Striking off is done with a straight 2x4. See p. 9.

haul-it-yourself concrete, it's terrific for medium-sized jobs. Check with your ready mix dealer. He may well have such a setup.

Trailer-mixers that are the basis of haul-it-yourself hold up to 1 cubic yard of concrete. Each has its own means of dumping. Since they're filled at a batch plant, you are relieved of buying and proportioning materials. Mixing too. The trailer drum rotates while you drive and does the mixing. (We don't recommend using haul-your-own-concrete in hopper-type trailers that have no mixing drums. The materials settle out quickly, resulting in poor concrete.) The trailers are easy to maneuver.

The cost of haul-it-yourself mix isn't much more than for ready mix. Usually a half hour of trailer time is included in the rental. Prices are higher on Saturdays when the mixers are especially popular. You'll also have to pay more if you return the trailer uncleaned.

One caution on buying haul-it-yourself concrete: buy the mix only from a reputable ready mix producer. A number of "investors" are getting into this business who know little about producing good concrete. Rent their mixer, if they'll let you, but get your concrete from a good ready mix man.

Haul-it-yourself concrete calls for at least two men, one to handle the mixer and another to handle the concrete. A crew of three would be better.

BUYING READY MIX

Professional concrete contractors don't

Striking 2x4 is zigzagged the over tops of the forms. Excess concrete is pushed off the edges. Screeding makes a flat, even surface. Surface should be floated twice with the wood float, once immediately after strikeoff and once after the water sheen disappears. Move the float in sweeping arcs. Follow floating with edging and jointing. The edger is drawn back and forth along the forms using them as a straightedge guide. Down-pressure should be very light. A 1″x12″ plank is laid down as a guide for cutting contraction joints. Note how Roth holds the leading edge of the jointer up off the surface to keep it from digging in.

mix their own concrete. They buy ready mix. You can too, if the size of your job warrants it.

The easy way to use ready mix is to tell your dealer what you want to do with the mix and let him choose the proportions of ingredients. Don't order by price alone, or you may get cheap concrete. Choose a dealer who is a member of the National Ready Mixed Concrete Association.

Figure up how much mix you need using the method described on page 43. Order 10 percent additional to allow for error and waste. Call your dealer and arrange when and where you need the mix.

Be ready before the truck-mixer arrives. Have your crew ready too. You'll need a minimum of three men.

If the weather is hot and dry, cut your day's work plans in half. Don't pour at all when the temperature is below 55°F.

Most truck-mixers can reach up to 12 feet with a pour. If you must reach farther, build a wooden chute out of a 2x12 plank with 1x6 sides nailed on. Or rent a metal chute from your dealer. If chuting is out of the question, use a rubber-tired wheelbarrow to cart the mix in and dump it.

Put planks across the sidewalk, driveway, over sewer lines and on your lawn wherever the truck-mixer will drive. These are to prevent serious damage from the heavily loaded truck.

TOOLS

You'll need the following minimum concrete tools for easy working: shovel,

Steel-troweled finish is tough to get. If you try it, move the float in wide arcs over the surface raising the leading edge slightly to avoid dig-in, as Woods demonstrates here.

straightedge, wood float and steel trowel. Also very useful are: lightweight metal float, edger, jointer-groover and kneeboards.

A flat-ended shovel lets you work the mix easiest. The straightedge need be only a really straight 2x4 long enough to reach across the forms and stick out about 6 inches on both sides. A straightedge is also called a *screed*.

You can make your own wood float too out of a piece of a 1x4 redwood or pine board 15 inches long. Screw a wood handle onto it, recessing the screws in the float's bottom. The steel trowel is used for making very smooth finishes on slabs. A metal float is like a wood one, but it puts a smoother finish on the surface.

An edger makes neat, rounded edges on slabs and tops of walls. The jointer-groover cuts contraction joints in the slab and should have a 1-inch bit to make a groove that's at least one-sixth the depth of the slab.

Kneeboards let you get onto the surface of a slab to finish the parts of it you cannot reach from the edges. You can make a pair from two 18-inch-square pieces of ⅜-inch exterior plywood with 1x2 or 2x2 cleats to let you pick them up.

The easy way is the best way in finishing. Work the surface as little as possible. Follow this procedure to make a lightweight-metal-float finish for slabs. Steel-troweled finishes aren't easy. Lots of skill is required:

1. Place concrete against the forms slightly higher than needed.

2. Strike it off by see-sawing the straightedge over the tops of the forms.

3. Float lightly to smooth ridges and eliminate air pockets. The less the better.

4. When the water sheen disappears from the surface, float again with the wood float. Follow immediately by floating with a metal float (or steel trowel, if you must). Never float a surface with water standing on it. Squeegee the water off if necessary before you float.

5. Smooth all edges with an edger. Cut joints with a jointer-groover. Edge and joint by moving the tool back and forth along the forms or along a guidestrip of wood. Don't press the flat part of either tool into the surface, however.

6. When the water sheen disappears again, you may metal-float once more, (or steel-trowel). Edge and joint again. That's it.

Don't neglect the last step: curing. Every concrete project should be cured a week before you use it. Cure as soon as the surface can stand it to keep the concrete continually wet. You may do it by sprinkling with water or by covering with a sheet of polyethylene.

An easier method is to brush, roll or spray on a coat of W. R. Meadows Sealtight *Cure-Hard,* if you can find it. The coating will weather away after its job is done. This is called a *membrane* cure. It is usually sprayed on.

Walls are cured simply by leaving the forms on and covering the exposed top.

Whatever curing method you use, don't let the concrete dry out before the week is up or its strength will be affected adversely.

INTERESTING WALL TEXTURES

The masonry units and the way you lay them can create some unusual textures

IF YOU WANT something refreshing in masonry walls, use different bricks, blocks or lay them so that a texture is created. A good way to get ideas on what looks good is to tour your area and see what the professionals have done with textures. By emulating the design you like you can build your masonry wall project using the same treatment.

Also look at the unusual treatments pictured in this chapter and others. You may find what you like here.

Robert Cleveland

Want texture? Color? A screen wall? Your concrete products producer probably has a variety of blocks to suit. Screen blocks, split-blocks and other shapes are shown.

Robert Cleveland

Chimney concrete blocks set on edge make a different-looking wall. Other units that can be adapted to making walls are quarter-rounds, hollow standard and half-blocks.

Split concrete block is produced by breaking one large unit into two at the factory. It has a colorful, rough texture resembling some kinds of stone but at much less cost.

Bricks are not easy to lay this way, but it builds a screen wall with regular bricks. Vertical bricks are mortared then laid on top of temporary spacer bricks set up on edge.

11

LAYING MASONRY UNITS—1-2-3

Right height, plumb, level and all in a row, that's the mason's secret formula

YOU CAN LAY BRICKS, concrete blocks and stones almost as well as a professional any day by following the simple steps that ensure success. But you'll never learn to work as fast as a mason. Doing a good job takes time. It also takes raw guts to rip out a row of bricks you've just laid when you find them slightly out of position.

The start of a good masonry wall is a good footing. Cast it of concrete and get it level so that the courses of masonry units —bricks, blocks, stones—get a level start. More on footings accompanies the separate projects in following chapters.

MORTAR

Almost as important as the footing is the mortar you use. Mortar not only fills the gaps between units, it bonds the units together into a solid wall. The mortar is liable to be the weakest portion of the wall. It's the first to need maintenance in the form of tuckpointing.

Don't try to make your own mortar. Instead, buy ready-packaged mortar mix. All you have to do is add water to it and mix. One 80-pound sack will lay about 27 concrete blocks or 65 average bricks with ⅜-inch joints.

Mortar is unlike concrete in that you should add as much water as possible yet keeping it workable. By workable is meant easy to spread, sticks well to the masonry units and allows the units to be tapped into postion without undue effort or without sinking too low. Workable mortar is sticky enough to hang onto the end of a unit when held horizontally. A good workable consistency for laying bricks and some con-

The easiest way to make good mortar is with a ready-packaged mix. Some concrete products dealers sell waterproofed cardboard mixing boxes. Beats mixing it on the walk.

Easiest method of all for mortar-mixing is with a small electric 5-gal.-pail mixer. Coloring pigment is weighed out and added directly to the rotating mixer drum as shown.

Bricklaying is simple. First set the corner bricks in mortar and tap them down and to the right course height. Tap with a mason's hammer (above) or wooden trowel handle.

When the corner brick is in position, use the level vertically to get the brick plumb. Level, plumb and the right height are the keys to corner bricklaying. Check each time.

crete blocks contains more water than a good workable mortar for laying standard-size concrete blocks. The reason is that the big blocks rest on a smaller amount of mortar. Therefore, the mortar must be stiffer to keep from squishing out.

MIXING MORTAR

You can mix mortar by hand on the sidewalk, in a trough-like mortar box or in a wheelbarrow. Boxed mixing is easiest because the mortar can't get away from you. Mix with a hoe, preferably, or a shovel. First mix the whole bag of mortar. Then, if you're only going to use part of it, put the rest back in the bag. Dry-mixing of the whole bag is necessary because, like premixed gravel-mix and sand-mix, mortar-mix ingredients tend to settle out in the package.

Make a hole in the center of the pile and pour some of the premeasured water into it. The required water is given on the sack of premix mortar, usually 5½ quarts for 80 pounds of mix. Mix, then add more water and mix again. Finally add all the water and mix until all the ingredients are blended completely, just as for concrete.

Transfer the mortar to the mortarboard —an 18-inch square of plywood. Try lay-

ing some units with it. If too stiff, add more water and mix on the mortarboard with the trowel. In fact, any time you feel that the mortar would work easier if it were wetter, you may add water and mix it in.

Adding water to mortar that has stiffened on the mortarboard is called *re-tempering*. There's no need to throw this mortar away as long as it has not begun to set. Setting does not begin for about 2½ hours after mixing. Mortar that has not been used in 2½ hours and has parti-

NOMINAL SIZES OF BRICKS

BRICK TYPE	HEIGHT	THICK-NESS	LENGTH	NUMBER PER SQ. FT. OF WALL (Allow 5% waste)	SACKS OF MORTAR MIX PER 100 SQ. FT. OF WALL (Allows 10% waste) Joint Thickness 3/8 in.	1/2 in.
Modular	2⅔"	4"	8"	7.4	6.6	8.3
Engineer	3⅕"	4"	8"	6.2	5.8	7.3
Economy	4"	4"	8"	5.0	5.0	6.3
Double	5⅓"	4"	8"	3.7	4.1	5.2
Roman	2"	4"	12"	6.6	8.0	11.0
Norman	2⅔"	4"	12"	5.0	6.1	7.8
Norwegian	3⅕"	4"	12"	4.1	5.2	6.7
King Norman	4"	4"	12"	3.3	4.4	5.6
Triple	5⅓"	4"	12"	2.5	3.6	4.6
SCR Brick	2⅔"	6"	12"	5.0	9.4	12.2

STANDARD RUNNING BOND

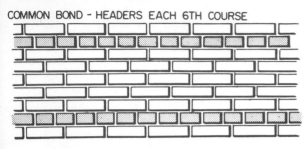

COMMON BOND - HEADERS EACH 6TH COURSE

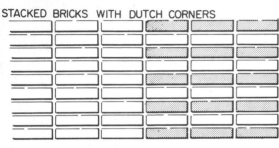

STACKED BRICKS WITH DUTCH CORNERS

Laying a row of bricks from the corners lets you build the corners several bricks high. Allow time to set before stretching string. Level is used as a straightedge.

ally set should not be retempered but thrown away.

You'll find that half your time is spent in mixing mortar if you do it by hand. If the project is a big one, get a concrete mixer to do the work for you. Best for mortar is the 5-gallon-pail mixer shown in the photographs. It will handle about one-third of a bag of mortar mix at once. This is about the amount you can hold on the mortarboard. If your wife or one of your kids can take over mortar-mixing, you don't have to worry about that part of the job. You can concentrate on laying the units. Machine-mixed mortar tends to be more workable too, because it gets worked more.

Colored mortar may be made by adding 1½ pounds of concrete coloring pigment to an 80-pound sack of premixed mortar before dry-mixing it. If a faint hint of a color is all that's desired, add less.

TOOLS FOR MASONRY

The same masonry tools work for laying bricks, blocks, and stones. They are: mason's trowel, level, rule, mason's hammer, stringline and blocks, jointing tool and blocking chisel.

The mason's trowel becomes an extension hand when laying masonry units. Pointed at one end, it's used for picking up, carrying, spreading and trimming mortar. There are many sizes of trowels, but the handiest for general use measures about 5x10 inches. The lighter the better.

The usual mason's level is 4 feet long and made of wood. If you don't have one, you can get by with a 2-foot wood or metal carpenter's level.

The level is used as level, plumb and straightedge. It positions footing forms, plumbs masonry corners, levels courses, levels individual units and checks alignment. Its body must be straight.

The rule can be a tape measure or, better yet, a 6-foot folding wood rule. Best is a mason's folding rule marked off in perfectly spaced courses. The rule is mostly used vertically for setting corner units to the right height.

A mason's hammer has a square head at one end and a chisel head at the other. It's used for scoring and breaking units to size and for tapping them into position.

A mason's stringline should be both

Butter the end of a brick fully with mortar before laying it against another. If the bricks are dampened and the mortar is workable, it will stick, even upside down if needed.

Stretch a string between the corner bricks, spread mortar and lay the in-between bricks to line. Push downward and toward the other brick at same time. Tap into position.

light and strong. This lets you pull it tight enough to eliminate center sag, thus getting level courses when the units are laid to the line. At each end, the line should have blocks that help hold it in position on the wall. Blocks make stringline easier to use.

A jointing tool—either rounded or V-shaped—is necessary to make neat concave or V-joints in walls. For brick the tool is short. For concrete blocks the tool should be at least 22 inches long. The extra length helps it smooth out irregularities in joints.

To make clean, square cuts in face bricks, you'll need a blocking chisel. It's like a cold chisel but much wider. The 4-inch blade spans over the face edge of the brick being broken. A blocking chisel is struck with a mason's hammer.

LAYING BRICKS

Bricks are laid in what's called a *full mortar bed*. Bed joints are the ones above and below the bricks in a wall. Bricks also need full head joints, that is, the joints between the ends should be full of mortar.

If possible your project should always be sized to come out even in full and half-bricks. This makes it look better. Easier to build too because there's less cutting of bricks.

The last brick in a course is called the closure brick. To lay it, butter both ends of the brick and the opening. Then lower the closure brick straight down to its bed.

To break bricks for filling odd-shaped spots, chisel a line across them with a mason's hammer. Chisel back and forth in a line until the brick breaks in two; it works fine.

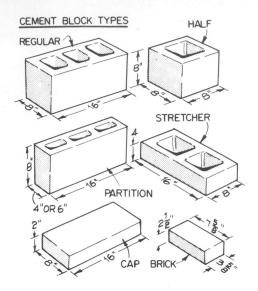

CEMENT BLOCK TYPES

REGULAR   HALF

STRETCHER

PARTITION

CAP   BRICK

Concrete blocks are laid in mortar beads spread along the outer and inner edges. As in brickwork, the corner units should overlap one another. Tap to level and height.

The procedure in bricklaying is as simple as this:

1. Lay the corner bricks to the right height, plumb above one another and level in both directions.

2. Stretch a stringline between the corners flush with the top edges of the corner bricks. Lay the in-between bricks with their tops flush with the line and 1/16-inch away from it. See that they're level across as you lay them.

Each brick should correspond to the one in the course below according to the *bond* or pattern you choose. The strongest and simplest pattern to lay is a running bond. In it, each brick is centered over the joint between the two bricks below it. Keep the pattern intact as you lay.

In practice, the corners are laid up three or four courses ahead of the center of the wall before stretching the stringline. This is to give the mortar beneath the first-used corner bricks time to set up so it won't be pulled out of place by the stringline. Laying up corners takes more skill than laying bricks to a stringline because you have to lay a row of several bricks without the aid of a line. This means that you must use the level as a level, plumb and straightedge.

Joints in brick walls may be from ¼ to 1 inch thick. Easiest to lay are ½-inch joints. The wider the joint, the more mortar-like the wall will be. Wide joints, however, help cover up goofs in bricks and in bricklaying.

If the wall is more than one brick wide,

you'll need more than one tier of bricks. Build them up at the same time, spreading mortar solidly between the two tiers. The end joints in bricks that are in the same course but in opposite tiers should be staggered.

The tiers of bricks must be held together. This is done by placing *headers,* bricks laid cross-wise of the wall, over both tiers. A minimum of 4 percent of the bricks in a wall should be headers. Maximum distance between headers is 2 feet horizontally and vertically. Lacking a pattern that permits the use of headers in a two-tiered wall, you may use masonry ties—crimped pieces of galvanized metal—about three feet apart in every course. There should be a tie for every 4½ square feet of wall surface. Max. vertical distance between ties is 3 ft.

The last brick in a course is called the *closure* brick. Lay it by buttering the opening and the ends of the brick before laying.

If when laying a brick, it sinks out of position below where it should be, pull it out, scrape up all the mortar around it and relay it in fresh mortar.

Keep building up corners and laying in-between bricks until the wall reaches full height. Then you'll probably want to cap it with bricks set on edge, stones or blocks.

Bricks should be throughly dampened with water before laying them. This keeps them from sipping up water from the mortar and destroying the bond. Soak the pile with a hose before you start bricklaying for the day.

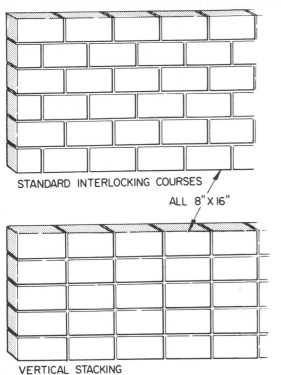

STANDARD INTERLOCKING COURSES

ALL 8"×16"

VERTICAL STACKING

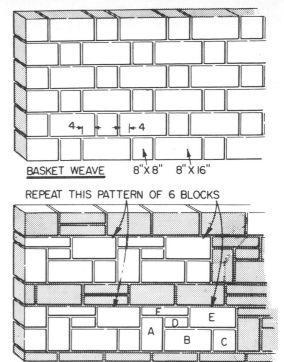

BASKET WEAVE 8"×8" 8"×16"

REPEAT THIS PATTERN OF 6 BLOCKS

PATTERNED ASHLAND – A– 8"×12"
B&E– 8"×16" C– 8"×8" D– 4"×8" F–4"×16"

LAYING BLOCKS

On the other hand, concrete blocks should never be wetted before laying. In fact, they should be covered to keep them dry. Blocks laid wet will later shrink, causing cracks in the wall.

Blocks are laid like bricks, except they get what's called *face shell bedding*. Block mortar is spread on both face shells and in two rows at the edges of the end joints. Only the bottom row of blocks is laid in a full mortar bed.

Blocks are designed for ⅜-inch joints. Thus a standard 8x8x16-inch concrete block actually measures ⅜-inch on a side less than its nominal dimensions.

Because of their large size, blocks lay up faster.

In blocks, the bottom row should be laid out dry without mortar to let you space out the row to come out in full and half-units. Mark footing for each and remove all.

Block wall corners should be laid one course ahead of in-between blocks because blocks are heavier and don't need so much setting time before you stretch a stringline across them. Professionals don't do it this way, but we should.

Don't mortar more than a few blocks ahead. The mortar stiffens fast.

LAYING STONES

Stones should be laid in full mortar, like bricks. Many different kinds—or "cuts" of stones are sold. Some are completely trimmed on all six sides.

Rubble masonry, made with rocks laid as they come, may be laid *coursed, uncoursed* or *random-coursed* depending on the appearance you want.

In buying rubble masonry it's okay to prop stones in place by putting chips of stone under them while the mortar sets.

Always set stratified stones such as slate on their natural beds, not on edge. Use the biggest stones toward the bottom of the wall, not at the top.

Stones are trimmed to fit the space for them with your mason's hammer.

Mortar needed in stonework varies from 15 to 35 percent of the wall's volume.

After laying a masonry wall, and tooling the joints, dry-brush all mortar stains from the face of the wall. On bricks and block walls follow with an acid-etch using a 10-percent solution of muriatic (hydrochloric) acid. Rinse with plenty of water to remove all traces of the acid. Try wire-brushing a stone wall to clean it, but don't use acid on stones. It stains them.

Party times around the barbecue grill are your reward for an enjoyable weekend making the grill. Bricks are set in ready packaged mortar mix. Wire grill surface top added.

YOU-BUILD-IT BARBECUES

A permanent outdoor barbecue is no "luxury" when you build it all yourself

HERE ARE THREE brick barbecue grills you can build without sweat. Although one is built in conjunction with a concrete patio, all need some sort of paving around them. If you already have the patio, your best bet is to add the barbecue beyond the edge of your patio at a convenient spot.

Construction of the two circular barbecues is similar. The low one takes less materials and is easier to build.

The higher one, which is combined with a bench, is more convenient to use. In it the actual barbecue is a steel unit. You might consider recycling your present portable barbecue by building it into a masonry one.

The rectangular barbecue is small and a snap to make. In spite of its diminutive size, it will impart the same zesty flavor to food as a more elaborate outdoor fireplace with chimney and all.

Gratings are made of ½-inch steel bars set 2 inches apart. The upper one is for food. The lower one holds a wood fire.

In between is a ledge of offset bricks on which you can place a pan of charcoal. The barbecue footing should be made of cast-in-place concrete. Reinforcing bars indicated in the drawing are optional.

The dimensions of the barbecue are not important, as they would be in a fireplace. The size of barbecue can and should be varied to suit the size of bricks you build it with.

One thing. Laying bricks vertically in circular soldier courses, as the pros call them, is more difficult than laying them flat. It calls for a good, sticky mortar that won't slough off when you hold the brick upright to lay it. After laying a brick, mortar is chunked in to fill the wide gap at the outer edge of the soldier course. Make sure that these courses are full of mortar. They show as very wide mortar joints on the outside of your barbecue.

You may, if you like, line the interior with special firebricks laid in fireclay. There's no need however. Ordinary clay bricks and mortar work fine. Let the mor-

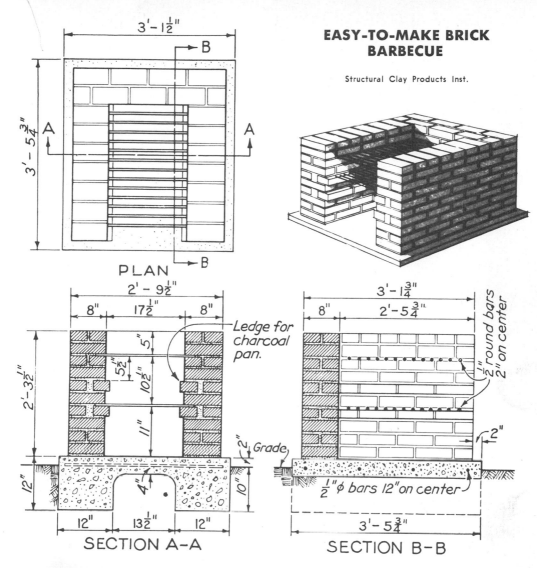

EASY-TO-MAKE BRICK BARBECUE

Structural Clay Products Inst.

PLAN

3'-1½"

3'-5¾"

B

A — A

SECTION A-A

2'-9½"

8" 17½" 8"

Ledge for charcoal pan.

2'-3½"

5"

5½"

10½"

11"

2" Grade

12" 10"

4"

12" 13½" 12"

SECTION B-B

3'-1¾"

8" 2'-5¾"

½" round bars 2" on center

2"

½" φ bars 12" on center

3'-5¾"

Barbecue-patio bench is made by laying bricks around a metal barbecue grill. Eight brick pilasters reach out 1½ brick-lengths to support the handy wood-topped bench.

Robert Cleveland

tar joints set a week or so before you use your grill. Using these techniques, you may find it interesting to design a barbecue or a small outdoor fireplace of your own. The decor of your patio and garden and the architecture of the house may suggest a particular shape or even period that you would like to incorporate.

19

Dig a round hole 2½ brick-lengths below finished patio (top, left). Install footing and gravel center. Pencil-and-string compass outlines for the base bricks on the footing. Set two soldier courses of bricks in mortar on the footing with face mortar between them (top, right). Set the first course to pencil mark. Pull out the wood drain-hole sticks. Now (middle, left) lay two rings of bricks on top of the soldier courses, overlapping onto the patio. Spread Sakrete mortar generously and tap each brick into it until flush. Here's how the upper course (middle, right) matches with the soldiered bricks below and the top of the patio. Install 4 eyebolts in mortar below the top bricks to hold a grill. Chunk the wide mortar joints (bottom, left) around the outside of your barbecue with mortar. Slice it off flush with the bricks, using your masonry trowel like a knife. Slice off the excess mortar (bottom, right) on top of the barbecue too. Note the ¾" air vent holes formed in the mortar joints between some bricks. There should be about six in all, two, as shown, on all four "sides" of wall.

STANDOUT MAILBOX STAND

This sturdy and attractive unit can be painted or even made in brick or stone

CARS THAT run off the road enough to damage this mailbox stand will likely be there the next morning. At the least the drivers will remember the offense. What's more, this mailbox stand is an asset to your property, more than a plain post would be.

Before building one, find out from your local building officials what the proper setback is. If you put the stand too close to the street, you may be liable for cars damaged by hitting it.

Dig for a concrete footing that's somewhat larger than the 16x18-inch concrete block stand. The footing, as always, should reach at least 12 inches below ground, deeper if frost penetration goes deeper. Make it 8 inches thick of cast-in-place concrete. There's no need to form for the edge. Let the earth excavation do that. Screed off the top level, however, even though it need not be very smooth.

When the footing has cured several days, you can lay the blocks. You'll need 12 6x8x16-inch blocks. Of these, 8 should be bullnose corner blocks. These make into a stand with gently rounded corners. The other blocks should be square-ended corner blocks. These are largely sold now as standard concrete blocks, especially in the West. At any rate, since they'll be laid vertically with their top surfaces exposed outside the mailbox stand, they should NOT be the type of blocks with ears at the ends. You want square ends. You'll also need 3 cap blocks to finish off the top of the stand.

Lay the blocks in mortar as shown in the drawing. Get them level in two directions and the corners straight and plumb.

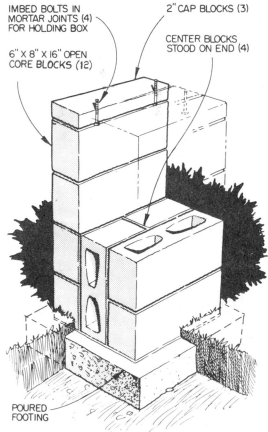

IMBED BOLTS IN MORTAR JOINTS (4) FOR HOLDING BOX

6" X 8" X 16" OPEN CORE BLOCKS (12)

2" CAP BLOCKS (3)

CENTER BLOCKS STOOD ON END (4)

POURED FOOTING

CONCRETE BLOCK MAILBOX STAND

Leave-in forms with alternate planter squares lead the eye toward the entry. The slabs were run with an edger and the redwood forms stained dark to stand out from concrete.

HOW TO MAKE SIMPLE WALKS

Garden walks, sidewalks, access walks, all make interesting easy-to-do projects

YOU NEED THREE types of walks around your house: garden, access and sidewalks.

Usually the sidewalk is provided by the city or town where you live. If you don't have a sidewalk across your lot and would like to build one yourself, consult the local government about its design requirements. You'll need to know things like elevation, set-back, width, thickness, finish, jointing and the like. These requirements may be standardized. If you build a sidewalk and get one or more of them wrong, the city may later tear out your walk and charge you for building another to specs.

An access walk has a different purpose from a sidewalk. It goes between the sidewalk and your front door, back door, into the back yard, to the trash disposal area, and to the garage. Building them is entirely your responsibility. Minimum access walk width should be 2 feet, this for service walks. The access walk to your front door should be at least 3 feet wide. Normally, concrete walks are a nominal 4 inches thick. The thickness should be increased to 5 or 6 inches where trucks will drive across them. At those points they should otherwise be designed as driveways too. See the chapter on building driveways.

As with other concrete that abuts an existing wall or slab, an isolation joint should be provided between the new and old concrete.

CONCRETE WALKS

Concrete access walks need a control joint at least every 4 to 5 feet along the walk. Joint depth should be ¾ to 1 inch. No reinforcement is needed in a sidewalk.

For drainage a walk should be finished with a crown—the center higher than the edges—or sloped from side to side, with one edge higher than the other.

A walk need not be a solid, continuous strip, as is so often seen in housing developments. With some thought and study of the photographs on these pages, you can do much better. Think of an access walk as a stretched-out patio.

Excellent access walks can be made of bricks, flagstones or precast concrete stepping stones. The paver units may be set directly on prepared ground, as is shown in the chapter on building a brick patio. Joints are mortarless.

Combination exposed aggregate and concrete tile access walk also is an edge for the lawn and flowerbed. Exposed-ag and tile base were cast; then the tiles were laid.

Wood forms of 2x4's staked at 4' intervals are best for building walks. Spread a leveling layer of sand, dampen and tamp it down. Cast one or more sections at a time.

Portland Cement Assn. photo

Drive 1x2 or 2x2 stakes to hold the forms in place. Arrange them so they slope about ⅛" per foot away from the house. Drive the stake below the tops of the forms and nail.

23

CONCRETE WELCOME MAT

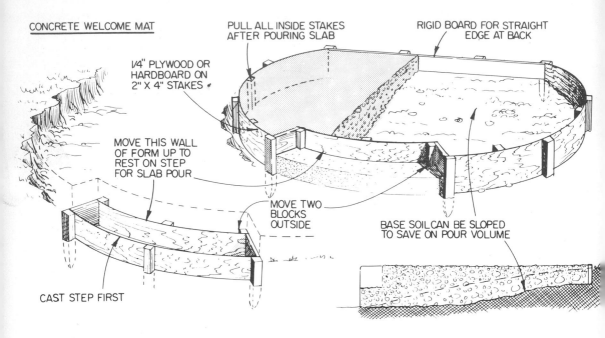

PULL ALL INSIDE STAKES AFTER POURING SLAB

RIGID BOARD FOR STRAIGHT EDGE AT BACK

1/4" PLYWOOD OR HARDBOARD ON 2" X 4" STAKES

MOVE THIS WALL OF FORM UP TO REST ON STEP FOR SLAB POUR

MOVE TWO BLOCKS OUTSIDE

BASE SOIL CAN BE SLOPED TO SAVE ON POUR VOLUME

CAST STEP FIRST

An exposed-ag concrete welcome pad breaks up an ordinary access walk and provides a step up to a sloping lot. It is cast in two phases: first the step, then the pad.

A series of outsized stepping stones between the sidewalk and house entry make a super cast-in-place access walk. Alternate circles are cast first, then in-between ones.

Robert Cleveland photos

Units for a higher-type patio are laid on a sand-cement subbase. Joints are mortared. The process is shown in the accompanying how-to photos.

WELCOME PAD

Even though you already have a plain ribbon-type concrete access walk in front of your home, you can dress it up with a simply built concrete "welcome mat." The exposed-aggregate finish shown in the accompanying photo can be imparted or you can choose some other finish.

The front portion of your existing access walk will have to be broken up and removed.

The welcome mat is built in two phases. The first phase is building the small concrete step in front. The step is not needed if the raise in grade will be 7 inches or less. If it's more than 7 inches, form and cast the step. It should be half the thickness of the welcome pad, but not higher than 7 inches. The tread should be 12 inches wide.

The second phase is to form and case the mat itself. Its front portion fits around the already-built step. Form the curves with ¼-inch plywood.

The welcome mat need be only 4 inches thick. If it must appear thicker in front because of a grade change as in the photo, form the edges full-thick. Then pile broken-up pieces of your access walk or rocks around in the center of the forms. This will reduce the pad's thickness through the center and save on concrete required. Be sure that no portion of the mat will be less than 3⅝ inches thick, however.

When the second half of the pad has been cast and the forms removed, the welcome mat should look pretty much like the one in the photograph. Of course, you can vary its shape, size and appearance: It could be square, rectangular or any other shape you like.

CIRCLE WALK

A variation on the welcome mat is the circle access walk. Use it in place of your entire ribbon-type walk. Cast alternate circles first, varying their size and possibly their finishes for added interest. Use the

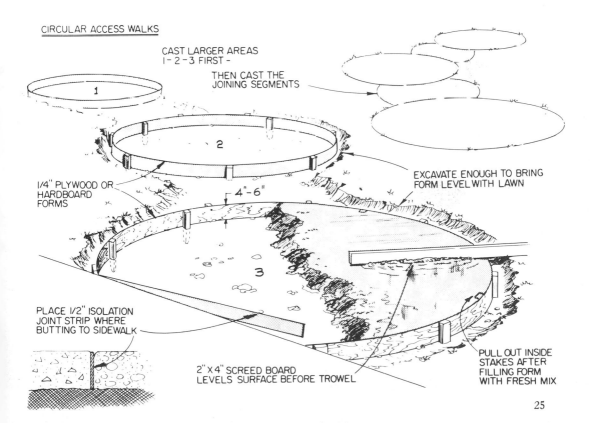

CIRCULAR ACCESS WALKS

CAST LARGER AREAS 1-2-3 FIRST -
THEN CAST THE JOINING SEGMENTS

1

2

¼" PLYWOOD OR HARDBOARD FORMS

4"-6"

3

EXCAVATE ENOUGH TO BRING FORM LEVEL WITH LAWN

PLACE ½" ISOLATION JOINT STRIP WHERE BUTTING TO SIDEWALK

2" X 4" SCREED BOARD LEVELS SURFACE BEFORE TROWEL

PULL OUT INSIDE STAKES AFTER FILLING FORM WITH FRESH MIX

Mortared brick access walk is a higher-type paving than the mortarless brick illustrated in the brick patio chapter. Use it when you want a deluxe job at a higher cost. It'll be worth it.

first-cast circles as partial forms for the in-between ones.

Grade changes along the way are easily handled by forming thicker slabs with broken rubble fill inside the forms.

Vary the shapes to squares, circles with squares or whatever grabs you.

The large shapes give the same effect as stepping stones, but on a massive scale, almost creating a patio between the sidewalk and your front entry.

STEPPING STONES

Small-scale stepping stones, the kind you can buy at products outlets or precast yourself and set in place, make excellent access walks. They're especially good as service walks and garden paths. As you might expect, the cost is small. Not only is less square footage of surface needed, the step stones need be only an inch or so thick. This means that much less concrete is consumed by stepping stones than a regular poured concrete walk.

Besides using precast stepping stones, you can dig holes and cast them in place.

To cast in-place stones, dig out the sod and soft soil. Fill if necessary with gravel tamped down hard. The holes should be deep enough that the tiles can be about 1½ inches thick.

Setting precast stepping stones is done in dug holes. Get the bottom as flat as possible. Then provide a layer of sand-mix concrete placed dry as a leveling course over the soil. Place the step stone on that, wiggling it into firm contact. Replace the sod close around each tile. The sand-mix layer will harden as it takes on moisture from the soil and provide uniform support for your tiles. It helps keep them from breaking.

CAST YOUR OWN

If you cast stepping stones yourself, make your own hinged or clamp-type take-apart form to ease the job. Make it about 1½ inches high so the tiles produced will be that thick. Then the tiles may be as large as about 3 feet square without danger of cracking.

When you must make enough step

High-quality mortared brick walk is made with a unique sand-cement subbase. Spread dry Sakrete sand-mix concrete in an edge trench. Set 1x3 form 5″ above the trench.

Spread a workable Sakrete mortar mix along the edge next to the form. Tap a row of bricks on edge until their tops are flush with the top of the form. Then add mortar.

Sakrete sand-mix makes a great subbase. Dump it between the rows of edge bricks just as it comes from the bag. On its first wetting the subbase will harden into a slab.

Make a wood spreader to screed the subbase off exactly one brick thickness below the edges. Work it into the corners too to get a flat surface for placement of bricks.

Lay the paving bricks in the desired pattern with about ½″ spaces between them. Once you get it right so that the pattern works out, use your fingers to gauge the spacing.

Chunk mortar into the joints, working it down with a small bricklayer's trowel until each joint has been completely filled with mortar. Clean up the excess mortar later on.

27

Robert Cleveland

Real flagstones can be set in a sand-mix concrete subbase or on a concrete slab in mortar to produce this attractive walk. Black mortar matches stones' color nicely.

Robert Cleveland

Rectangles cast parallel with the house but each one offset from the other makes a walk that angles across the yard to a side entrance. Forms and concrete are painted.

Stepping stones leading to an exposed aggregate slab entryway have a rock salt finish to contrast with that of the entry. These could be either cast in place or precast.

Robert Cleveland

EASY HINGE – APART FORM FOR STEPPING STONES

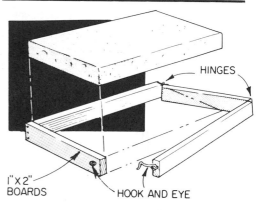

1"X 2" BOARDS

HINGES

HOOK AND EYE

stones to build a very long walk, it helps to be able to work production-line style. This lets you build one tile after another in a single form. The secret is to use a stiff mix. It will stand up with little edge-slumping when the form is removed immediately after finishing the tile. Cast tiles on polyethylene sheeting on a flat surface with room to make all of your tiles. Being

gentle in form removal will help to preserve the embryonic tiles.

TEXTURE AND COLOR

It is only a little more effort and expense to put good looking texture and color into the top surface of your precast tiles. See the chapter covering this. Any of the fin-

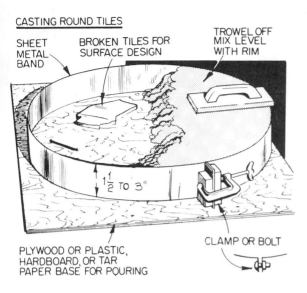

CASTING ROUND TILES

SHEET METAL BAND — BROKEN TILES FOR SURFACE DESIGN — TROWEL OFF MIX LEVEL WITH RIM

1½ TO 3"

PLYWOOD OR PLASTIC, HARDBOARD, OR TAR PAPER BASE FOR POURING

CLAMP OR BOLT

Can't keep people from taking a short-cut across your lawn to reach the door! Join 'em. Place a string of step-stones to keep the foot traffic from killing off your grass.

Mixture of textures, garden walk and stepping stones can be linked with concrete that has a different finish from the cast-in-place stones. Use metal forms for stones.
Robert Cleveland

Grade changes in a walk can be made with separate slabs. Framing can be left in for effect. Note angular pattern. Straight lines make for easy installation of concrete slabs.

ishing and coloring methods will work for making stepping stones, just as for slabs and castings.

Exposed aggregate goes especially well on tiles. Many of those you can buy have this finish. Color, of course, is often appropriate if it isn't too garish. Tooled-in patterns, brooming and cast-in patterns are also excellent textures.

A small 5-gallon-pail concrete mixer is a handy size for making stepping stone concrete.

One word of warning: don't make your tiles so thick and heavy that you won't be able to set them by hand. A good maximum size makes use of one sack of ready-packaged concrete mix. This ensures a tile that you can handle.

Portland Cement Assn.

The completed table before backfilling and landscaping looks as indestructible as it is. The plain-gray appearance of table can be enhanced by coloring. See chapter on Coloring, p. 90.

QUICK PICNIC TABLE

Want a picnic table that needs absolutely no maintenance and yet will last a lifetime? Here's how to make one

ARE YOU TIRED of refinishing the old wood picnic table? Legs getting a little rotten where they contact the ground? Then treat your family to a permanently installed concrete picnic table.

It is precast in wood forms made from 2x4's that have been ripped to 2¾" wide. Once you've made the forms, you can cast as many tables as you like. The form makes all the parts for a table at one casting. If you like, you can make a table every weekend and have tables for friends or to sell.

The drawing shows how the table goes together, how the form is assembled and how to cut the various parts of the form. Only four different designs are used for all form pieces: A, B, C and D. You will

need two of each. The forms are drilled to receive steel reinforcing rods. Placement of the rods is shown by the dotted lines in the drawing.

ASSEMBLY HOLES

Holes for assembling the top to the legs are made by greased bolts stuck into the freshly poured mix. The bolts fit through 2x4 templates that position them accurately according to locations shown on the plan. When the template is removed and the parts of the table separated, the greased bolts are pulled out. Protruding rebars in the table and bench leg sections slip into the bolt holes in the table and bench tops.

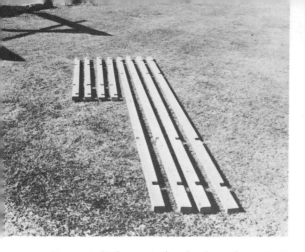

Cut out all the parts for the form from 2x3's or 2x4's ripped down to 2³/₄ inches. The best way to cut the notches is to make parallel cuts and break out with a chisel.

The notches permit half-lap joints in which one form member holds the other in position without nails or other fasteners. Notches should be 1⁵/₈ inches deep. See art, p. 32.

No. 3 (³/₈" dia.) reinforcing bars are used to strengthen the concrete. Insert them in the drilled holes and wire together. Set form on plywood covered with a bond-breaker.

Heavily greased bolts are positioned in the table and bench top parts with two 2x4 templates. Bolts are removed after concrete has set and the forms have been stripped.

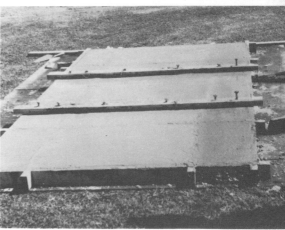

A disadvantage of this picnic table is that it cannot be moved. The legs are cast into footings below ground. Once placed, you'd need a sledge hammer to get it out. That's its benefit, too. The kids who use the table need the same determination to wreck it.

COLORING

Your table may be left plain gray, it can be cast with colored concrete, or an exposed aggregate finish can be applied

during casting, all to get color. It could be stained just as well after erection, or painted with non-dusting masonry paint.

A terrific terrazzo effect can be gotten by embedding tiny marble chips in the surface of an integrally colored mix. Then the next day you could take a disc sander to the surface of all the castings and grind away the mortar covering exposing the colorful marble chips. It would end up almost as smooth a surface as real terrazzo but would not require wet-grinding with a special machine. Later when the mix

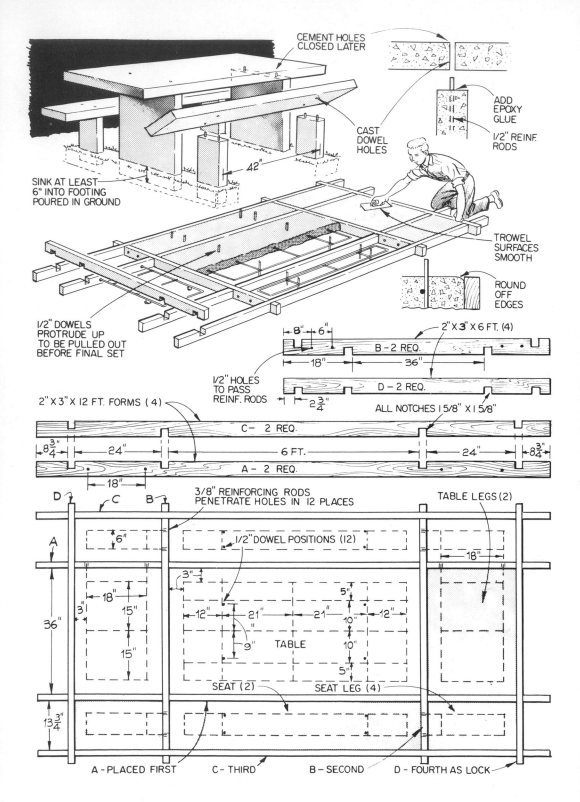

CEMENT HOLES CLOSED LATER

ADD EPOXY GLUE

1/2" REINF. RODS

CAST DOWEL HOLES

42"

SINK AT LEAST 6" INTO FOOTING POURED IN GROUND

TROWEL SURFACES SMOOTH

ROUND OFF EDGES

1/2" DOWELS PROTRUDE UP TO BE PULLED OUT BEFORE FINAL SET

1/2" HOLES TO PASS REINF. RODS

8" 6"

2" X 3" X 6 FT. (4)

B - 2 REQ.

18"   36"

D - 2 REQ.

2 3/4"

ALL NOTCHES 1 5/8" X 1 5/8"

2" X 3" X 12 FT. FORMS (4)

C - 2 REQ.

8 3/4"   24"   6 FT.   24"   8 3/4"

A - 2 REQ.

18"

D   C   B

3/8" REINFORCING RODS PENETRATE HOLES IN 12 PLACES

TABLE LEGS (2)

A

6"

1/2" DOWEL POSITIONS (12)

18"

3"

18"

5"

36"

3"

15"

12"   21"   21"   12"

10"

15"

9"

TABLE

10"

5"

SEAT (2)

SEAT LEG (4)

13 3/4"

A - PLACED FIRST   C - THIRD   B - SECOND   D - FOURTH AS LOCK

32

Here is how the picnic table parts look when the forms have been removed. Edging each slab with a rounded concrete edger gives a protected finished look to the pieces.

Cast the table and bench legs in concrete footings and place the tops over protruding rebars above the leg sections. Check to be sure everything is perfectly level and even.

Holes for assembly are grouted over with a 1:3 cement/sand mix. Slice it off flush with the top of the table. Do the same for the two bench tops. Cement will dry same color.

hardens enough to strip the forms—three days should do it—you could disc-sand the slab edges. They'd be fairly hard to expose at this point, but there would not be much surface to work on, so it could be done.

Painting a coat of concrete form retarder onto the forms before casting would make edge-grinding go quicker. (See the chapter on special finishes and effects.)

If you want an outstanding picnic table, try the terrazzo route.

Clean off and dry out the forms for future use. You will have lots of requests to borrow them—though you may want to cast a few extra tables yourself and then sell them.

BRICK PATIOS GO QUICKLY

Bricks need not be limited to the building of walls, they can just as successfully be used as paving

A BRICK PATIO, sidewalk or even a driveway isn't at all difficult to make. The idea is that the units are so small the work is light. Even a lady can do it. Anyone can make a brick patio if he follows directions and doesn't omit a step. Most problems come from neglecting something.

Buy bricks that can take the rugged exposure. Contact with the ground in freezing and thawing temperatures requires the very best in bricks. When you buy your bricks make sure that you get hard-burned ones if temperatures in your area ever go below freezing. Ordinary bricks would soon break up under the strain.

In any kind of climate you want to get bricks that are not so porous that they soak up water and become stained. Low-cost porous bricks, often called *common bricks*, are no darned good as pavers.

Concrete block pavers make a good patio too. While they're heavier to handle, they lay up faster because of their larger unit area. Use same methods described for the bricks.

Portland Cement Assn.

Choose a well drained site. Outline your patio with stakes and string. No need to size it exactly to the dimensions bricks will lay to. It works out because bricks are small.

Dig out all grass and roots in the area to be paved. Dig ½" to 1" deeper than the bricks are thick. Discard most of the diggings because they really won't be needed again.

Frame your patio on three sides with bricks set on edge in a 2" ditch. Tap them in by hammering on a block of wood. Lay the fourth edge after setting all the pavers.

Now add enough sand to the dug-out area to make the patio bricks, when they are laid, even with the edge bricks. In the next photo you'll see how simple it is to get it even.

Cut a notched sand screed out of a 1"x4". It's lower edge should ride even with the bottom of the pavers when laid. Notched ends slide over tops of the edging bricks.

Once the sand bedding has been leveled to grade, you may want to roll out a layer of 15-lb. asphalt felt to keep weeds and grass from sprouting up through your neat patio.

35

Now comes the fun part. Lay down brick pavers, beginning in one corner and following your chosen pattern. Lay them tight, no spaces between bricks. This goes fast.

As an alternate method of edging you can set edgers flat, on sand, tapping them in with a hammer handle. The stringline here is used much as in bricklaying, to keep edge straight.

The sand screed is then run over the flat edge bricks to level your sand embedment layer for paving. See-saw the screed back and forth as you pull it forward with sand.

Lay the bricks on the sand. Note that this method builds a little patio at a time to let you work on the bricks, keeping the sand course undisturbed by un-fancy footwork.

RUNNING BOND PATTERN

You are not restricted to using only bricks intended just for paving. With some patio patterns, though, you'll need paving bricks because they are exactly twice as long as they are wide. However, a running bond paving pattern is practical with any face bricks that don't have holes in them.

Brick pavers very from 1½ to 2 in. thick, depending on the manufacturer. Face sizes are 4x8 inches, or a bit less.

Of course, brick paving may be laid on a concrete base and with mortar joints be-

tween the bricks. This makes a deluxe patio. You can do it this way if you wish. In that case use a 1-inch-thick mortar bed placed over a 3-inch concrete base in place of sand embedment. The procedure is shown in the chapter on building a swimming pool. There it's used for the pool deck.

The easiest method of all is mortarless. Sand is used as a leveling course and the bricks are laid on it. A mortarless brick patio is easy to build, economical and, if properly done, it's a permanent improvement to your house and grounds.

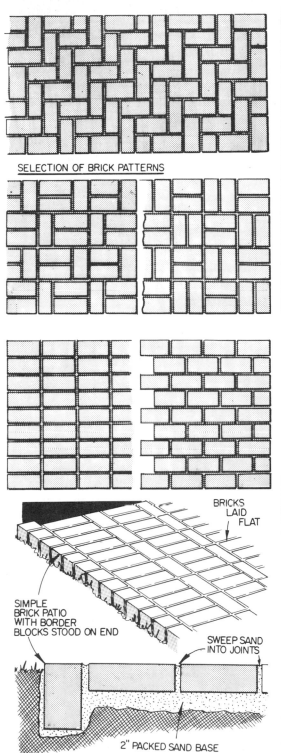

SELECTION OF BRICK PATTERNS

The final step in making a mortarless brick patio, sidewalk or driveway, is to spread fine mortar sand on the bricks and sweep it into all the joints between bricks until filled.

The completed patio is now edged all around and excess excavation is backfilled and planted with grass seed. You can then begin using your brick patio immediately.

Brick & Tile Service

BRICKS LAID FLAT

SIMPLE BRICK PATIO WITH BORDER BLOCKS STOOD ON END

SWEEP SAND INTO JOINTS

2" PACKED SAND BASE

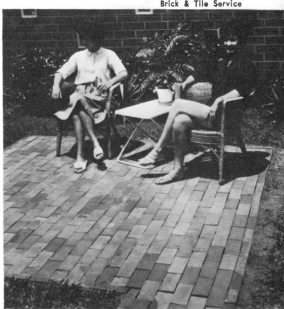

Rugged looking patio/pool deck uses wide gravel strips between slabs where the forms were. Two large concrete chairs were cast with same mix and design as used on the patio.

EASY-TO-MAKE CONCRETE PATIOS

Make your concrete patio one section at a time for absolute ease of building

IF YOU PLAN it right, building a con-crete patio needn't be a major project. At the same time you make things easier for yourself, you create a patio that will stand out from most others.

First of all, bring the patio down to your terms. Section it into workable-sized slabs with leave-in form boards. Then the patio can be cast in small pieces. Do as many pieces as you feel like and quit. If friends come to visit and offer help, take them up on it and do more pieces than usual. Eventually, the whole patio gets done. And leaving the forms in place adds to its design.

TREATED FORMS

Leave-in forms are best made of a rot-resistant wood such as all-heart redwood, cedar or cypress. However, if you'll soak ordinary Doug fir boards in creosote or penta wood preservative before building

them into forms, they will last reasonably well. I did this 10 years ago, and only now are some of the fir boards beginning to show problems. In a wet climate too. In a dry climate they'd likely last longer, if termites didn't get them. When the forms give up, you can fill the joints with col-ored mortar to look just as the forms did.

BUILD ON-GROUND

Another patio-taming trick is to forget the customary gravel, crushed stone or sand subbase. It's not needed anyway. In most locations you can build your patio right on the ground. This saves digging and it saves buying, hauling and placing of subbase material. The only spots where granular subbases are needed are in soft, mucky spots or those with poor drainage.

A snap-to-make patio will not have a smooth-troweled finish either. Forget that. It's for cement finishers and contractors.

The first step is to decide where you want the patio and to dig out all sod, roots, rocks and debris from the area. Dig to approximate depth of the patio concrete planned.

Ladder legs were nailed to long runners and laid on the subgrade. Then we nailed other long runners to the legs. A sledge hammer backs up the nailing to protect the forms.

Level around the edges of the forms and nail them securely to stakes driven into the ground. Good level is important to prevent puddles on your completed patio surface.

Frame the patio. Author helped neighbor Tom Miller build a patio over a dirt area in front of an outdoor fireplace. Miller chose thin 1½-inch section framing for the patio.

Level the center by stretching a stringline over three equal-sized wood blocks, two at ends, one to test with. Test block should just about touch stringline at each form.

Make a notched screed to level the sub-grade and subbase. Miller used sand only to fill in low areas and save on concrete. It was not needed for drainage in this case.

Any fill, even sand, should be compacted thoroughly to give uniform support to concrete placed on it. Miller made a workable tamper from a post and flat piece of wood.

With leave-in forms there's always a chance of the form lifting or dropping. The cure is to drive nails or drill and insert short wires where they will be cast into the slab.

With less skill and effort you can produce a much better looking finish with exposed aggregate or outdoor terrazzo, which is a form of exposed aggregate. For how to make them, see the chapter on special finishes and effects.

TEXTURED FINISH

Instead of exposed ag, you can save work by making a wood-float finish. One way is to wood-float but stop short of steel-troweling. This makes a somewhat rough patio and I recommend going further with the process. Follow wood-floating by floating with an aluminum or sponge rubber float. Both these produce good slip-resistant textures easily.

Brooming or brushing is another way to get a slip-resistant texture without sweat.

Make your grids a maximum of about 25 square feet, the smaller, the more manageable. Also, the smaller and more compact (squarish) they are, the thinner they can be. A thin patio saves on concrete, mixing labor and handling effort. For instance, a 4-inch-thick concrete patio can be jointed as far apart as every 10 feet. But if you drop the slab size to, say 4 feet square, you can reduce the slab thickness to 3 inches. If you continue the slab size reduction down to about 2-foot-square slabs, the thickness can be cut to 1½ inches. So make the slabs small and make your job easier.

Mix-it-yourself is good on concrete projects of a cubic yard or less. It gives great control over how much work you'll do in a day. You can rent the mixer from a local dealer.

Dump your mix into the forms from a wheelbarrow or wagon used to haul it from the mixer. In small slabs you can spread it with the float. Do one section at a time.

If you use ready-mix choose a convenient discharge spot. Put plywood under the wheelbarrow to protect the driveway from concrete spills. Care now saves work later.

Float the surface of the concrete to smooth and compact it. When placing thin slabs, the mix should not be overwatered because every bit of strength is needed for success.

PATIO SIZE

A common quotient for patio size is one-fifth the area of the house. Thus a 2000-square-foot house would have a 400-square-foot patio, or one measuring on the order of 20x20 feet. A snap-to-make patio, however, should be kept small. You can get by with one only 12x12 feet if it's well arranged. A large patio can be extended to take in the porch or pool deck. The patio should have access to the house as well as to the rest of the yard. A southern orientation will give you a longer season of use. Avoid a west-facing patio where you must stare at the afternoon sun.

The patio should slope away from the house ⅛ inch per foot in order to drain well. Avoid having low spots or "birdbaths" to collect water.

DIGGING

To start a concrete patio, dig out all the sod and black dirt in the patio area and a few inches beyond on all sides. The additional digging is needed to make room for setting forms. Level and compact the soil with a tamper. You can make the tamper from a 5-foot length of 4x4 with a 6x6-inch square of plywood nailed to its bottom.

Lay out the forms, usually 2x4's are used, and nail them securely to 1x2 or 2x2 wood stakes driven into the ground

A good way to cure a concrete patio is with the spray from a lawn sprinkler. Turn the sprinkler on often enough to keep the surface wet for 6 days. Try not to flood area.

Green squares are provided to relieve a large patio/pool deck. The wood forms were left in. Note how a long bench was used to enclose the patio along one side.

Robert Cleveland

Who says joints have to be parallel to the patio edge? The contraction joints in this interestingly shaped patio were cut at semi-diagonal angle dividing it into 4-ft. squares.

A plain concrete patio is enhanced by the serpentine brick wall laid around it. The wall also contains a built-in barbecue and dramatic effect night-time patio lighting.

Portland Cement Assn.

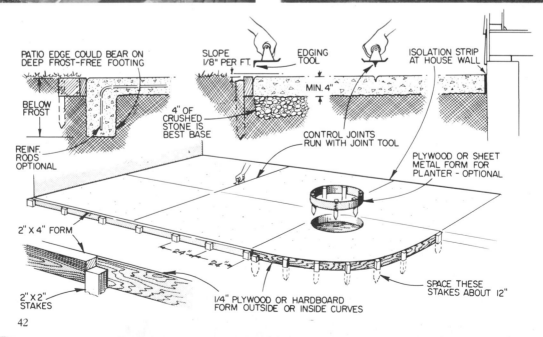

PATIO EDGE COULD BEAR ON DEEP FROST-FREE FOOTING

BELOW FROST

REINF. RODS OPTIONAL

SLOPE 1/8" PER FT.

EDGING TOOL

MIN. 4"

4" OF CRUSHED STONE IS BEST BASE

ISOLATION STRIP AT HOUSE WALL

CONTROL JOINTS RUN WITH JOINT TOOL

PLYWOOD OR SHEET METAL FORM FOR PLANTER - OPTIONAL

2" X 4" FORM

24" 24"

2" X 2" STAKES

1/4" PLYWOOD OR HARDBOARD FORM OUTSIDE OR INSIDE CURVES

SPACE THESE STAKES ABOUT 12"

at 4-foot maximum intervals. Use a level and stringline to help you get the forms at the right level and slope. Curving patio edges are harder to handle, but there's no reason you need to avoid them. Use ¼-inch plywood to form the curves. If the radius of the bend is less than 4 feet, be sure to cut the plywood so that the face grain is vertical and the plywood is bent across the grain. Dry bends as sharp as a 2-foot radius can be made in this way. Long-radius bends may be made with 1x4-inch lumber.

Lay out the curves with a garden hose. It's an old trick, but it works. Place the inside stakes along the curve and bend the form around them. Then drive the outside stakes and nail the form to them. You can then pull out the inside stakes and the form will be held in place by the outside ones.

Along curves, the form stakes should be spaced a maximum of 2 feet apart. Put one 2x2 stake at the joint between the curved and straight sections to hold them in good alignment at that point.

A thick, soft asphalt-impregnated felt material, called *isolation joint* should be placed between the patio and every existing slab or wall. Use it where the patio meets the house, sidewalk, porch or other. Lacking the ½x4-inch joint material, you can use a pair of ½x4 pieces of bevel siding. Arrange them so they nest into a rectangle between the patio and existing structure. Then after the new concrete

cures you can remove one, then the other. The joint left should be poured full of liquid asphalt or a nonhardening horizontal joint sealant to keep water out.

Before pouring your patio concrete, a clay soil should be soaked for several days. No standing water can be present, however.

CALCULATIONS

To find how much concrete you need, there are a number of formulas and tables around. I still prefer the old simple-arithmetic method. It works for any project. Here's how:

Write down all three patio dimensions in yards—length, width and thickness—and multiply them to get the number of cubic yards.

Suppose the patio is to measure 15x20 feet and be sectioned into small enough slabs that it need be only 2 inches thick. The width and length are given in feet, so divide each by 3 to put them in yards. The thickness is given in inches, so divide it by 36 to get it into yards. Then multiply.

You have $\frac{15}{3} \times \frac{20}{3} \times \frac{2}{36} =$ a shade less than 2 cubic yards of concrete.

It works for any concrete project.

JOINTING

When you finish the concrete, run an edger along all edges if you want them to have a rounded edge. At the least, all exposed edges should be run. Those against leave-in forms need not be run, and often look better if they are not edged.

If the forms don't do it, joint the patio into 10x10-foot maximum slabs. Otherwise it will crack when the concrete shrinks upon setting. To be effective, these joints must be at least one-sixth, better yet one-fourth, the depth of the slab.

You can leave openings in your patio wherever you want them. Simply form around them and place the opening next to a joint. This is because the opening creates a weak spot in the slab, where it will crack. The joint controls the crack and hides it.

Be sure to cure the patio slab concrete for at least six days before you use it much. From then on, your patio will be a lasting pleasure to the whole family. Outdoor living never looked better.

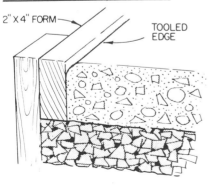

SECTION THROUGH A CONCRETE PATIO

2" X 4" FORM

TOOLED EDGE

Driveways can be reduced to one-day's-work slabs. Simply form up what you can handle in a day, perhaps a 10'x10' slab. Cast it, strip the forms and form up for the next slab.

WORK-SAVING TIPS

Here are a dozen or more ways to make your work easier without cutting down on the quality of the job—including some "easy" ways to do even better work

SOME OF the tasks involved in concrete-masonry building are just hard work. It doesn't hurt a bit to go about them in the simplest way, involving least effort.

In some things, the easy way isn't always the best. That can be true with concrete and masonry too. For instance, adding water to a concrete mix makes it much easier to place and finish. It's much easier to forget about curing than to apply one of the curing methods. Both of these easy ways are at the expense of quality.

Here, however, are some easy-way tips that will have no adverse effect on the quality of the job. Some of them can even help you to turn out a better project.

By making what's called a story pole, you save much meticulous measuring for brick course heights. Mark them off on a 1x2" stick. Then lay each corner brick to a mark.

When you mix concrete, position the mixer half-way between the piles of sand and gravel. You can load both into the mixer easily. Cement and water should also be nearby.

You can cover far more area faster when floating and troweling concrete if you rest one hand on an unused float. Marks left by the float are easily trowled away later.

If you brush off mortar stains from the brick faces at the end of each day's work, much effort will be saved when you finally acid-etch the wall to remove the mortar stains.

Moving concrete is hard work. Let gravity help. Ask your ready-mix producer to loan you an extension trough and use it to slide ready-mix into far corners of your project.

Rent the tools you need to save work on a big job such as building a patio. A gasoline-powered tamper makes quick work of compacting the soil under a future slab.

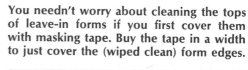

It may not seem very workmanlike, but a child's coaster wagon makes an excellent vehicle for handling excavated fill, sand, gravel, sacks of cement and other materials.

You needn't worry about cleaning the tops of leave-in forms if you first cover them with masking tape. Buy the tape in a width to just cover the (wiped clean) form edges.

If the forms are far apart, build a wet screed, as expert finisher Jerry Woods shows. Screed next to the forms, then do the center using the wet screed as a guide.

Steel-troweling is lots of work and it takes a great deal of skill. Avoid it by making a broomed finish. Float, then drag a fine broom lightly over the surface for texture.

An easy way to make a walk is with purchased precast squares and circles. They cost more than casting your own, but get you out of mixing, placing and finishing.
Robert Cleveland

If you're not fussy about laying a perfect brick wall, dispense with the stringline and use a long wood level to get straight rows. Rustic appearance can be quite attractive.

Award-winning driveway is a combination of squares and rectangles with leave-in forms. Black concrete defines parking. Pebble concrete leads to the front door. Marvelous effect.

DRIVEWAYS YOU CAN BUILD

This kind of project deserves special attention because

it is your welcome mat to guests and a setting for your home

YOUR DRIVEWAY deserves the best attention you can give it. Driveway-building can be made easier by using a thin section, minimum width and casting directly on the subgrade. Too many driveways are made 6 inches thick of sloppy, weak concrete when only 4 inches of good concrete would do a better job. A well made 4-inch driveway is ample for use by cars and light trucks. If heavier dual-wheeled delivery trucks will use the driveway occasionally, make it 5 inches thick. And if really heavy trucks like garbage and fuel delivery trucks will drive on it, better make the drive 6 inches thick and of quality concrete. All these are nominal thicknesses using 2x4- and 2x6-inch wood forms. A 5-inch driveway is made using 2x4 side forms and digging out one inch below the forms.

Of course, the thinner the section, the less work a drive is to build. In a 4-inch driveway, the slab is actually only 3⅝

inches thick, so take care that yours doesn't get built thinner than that at any point. If it does, it may be in trouble.

DRIVE WIDTH

The easiest driveway to build is a straight one of minimum 8-foot width. If there are any curves, the minimum width should be increased to at least 10 feet. If a driveway serves a two-car garage or carport, double its width.

A strip drive of two narrow ribbons of concrete might seem an easy one to make. Not so. It requires twice as much forming as a single drive. The strip driveway is still cheaper, though, because less concrete is required.

If the distance from house to street is great, you can build a single drive to within a couple of car lengths of the garage. At that point widen the drive to 18 to 20 feet. This lets a car swing into either stall.

Combining an entrance walk with your driveway will leave an unbroken front lawn as well as enhancing the driveway's appearance. Allow extra width for pedestrians to use the drive when cars are parked in it. A change in paving color or texture can help guide guests to the front door.

PROPER DRAINAGE

Maximum driveway slope according to FHA standards is 14 percent, or 1¾ inches of rise for each foot of horizontal distance. Grade changes should be gentle enough that cars won't bottom at front, center or rear.

Drainage is needed for a driveway, as for any large concrete slab. Most drive-ways slope enough from garage to street to avoid the need for cross-sloping. Drives that don't slope from end to end should slope from side to side. This can be handled in either of two ways. One side form may be set higher than the other, or the driveway may be crowned higher in the center than the edges. Either way, water will run off. The cross-slope, as it's called, should be on the order of ⅛ inch per foot of horizontal distance. A crown is imparted during screeding, by using a crowned screed rather than a straight one.

A driveway is simply a patio for your car. Many details of the construction are the same as for building a patio. A drive may be sectioned into small slabs with take-out or leave-in forms. These may be

A similar rectangular treatment is used to lead into a garage at an angle from the street. Both the exposed-ag and colored troweled-smooth concrete are used here.

The driveway and access walk of pebble concrete lead to an outdoor entrance foyer using the same paving. One-inch-wide forms left in, even at edges of the pavement.

A broad, curving driveway is a natural solution to reach from the street to a right-angle garage. To save on concrete, the drive could have been narrower at the street entrance.

Good slab sectioning makes many smaller pieces of driveway that are easy to build at the rate of one a day, working after hours. Exposed-ag at left is off limits to car traffic.

Robert Cleveland

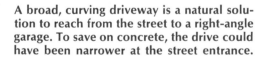

egg-crate style or any other form that doesn't leave long, slim projections that would be easily broken if overloaded from a car tire. Curves are handled the same way as for a patio. The driveway can flare from straight to curved, even make a full circle for turnaround.

Form staking should be the same as for a concrete patio. Once the forms are in place, check the subgrade depth by drawing a template over the form tops.

Like a patio, a driveway may be built directly on grade, but only if the soil is well drained. On poor-draining clay soil or where a drive must cross low land, use a 4-inch granular subbase. This may be crushed stone or gravel. It also acts as a

A great way to build a combination driveway-access walk uses pebble concrete for cars and plain concrete for people. It's formed with one pair of forms; see below.

POURING A DRIVEWAY BORDERED BY TWO WALKS

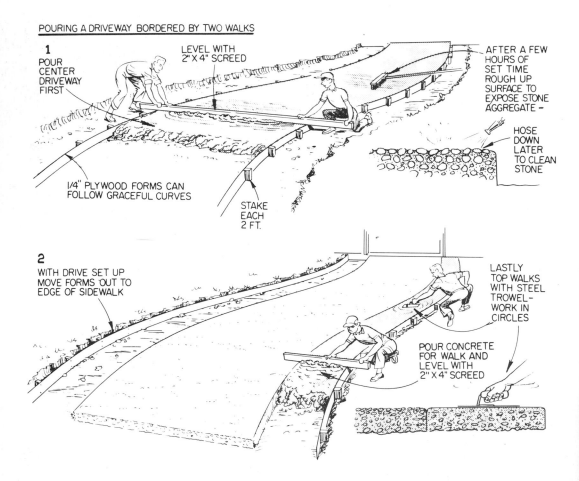

1
POUR CENTER DRIVEWAY FIRST

LEVEL WITH 2" X 4" SCREED

AFTER A FEW HOURS OF SET TIME ROUGH UP SURFACE TO EXPOSE STONE AGGREGATE –

HOSE DOWN LATER TO CLEAN STONE

1/4" PLYWOOD FORMS CAN FOLLOW GRACEFUL CURVES

STAKE EACH 2 FT.

2
WITH DRIVE SET UP MOVE FORMS OUT TO EDGE OF SIDEWALK

LASTLY TOP WALKS WITH STEEL TROWEL-WORK IN CIRCLES

POUR CONCRETE FOR WALK AND LEVEL WITH 2" X 4" SCREED

Color can do it for a driveway. Even if that color is black. It gives a sophisticated look and is merely a matter of adding black oxide or other coloring during the mixing.

Large pebbles give a cobblestone effect. Brick edging adds to the old-country impression this driveway gives to the house and grounds. Turnaround space is provided.

POURING A LONG CURVED DRIVEWAY

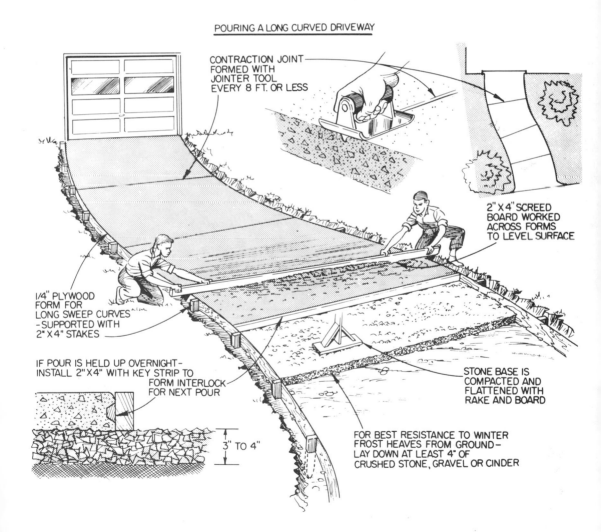

CONTRACTION JOINT FORMED WITH JOINTER TOOL EVERY 8 FT. OR LESS

2" X 4" SCREED BOARD WORKED ACROSS FORMS TO LEVEL SURFACE

1/4" PLYWOOD FORM FOR LONG SWEEP CURVES - SUPPORTED WITH 2" X 4" STAKES

IF POUR IS HELD UP OVERNIGHT- INSTALL 2" X 4" WITH KEY STRIP TO FORM INTERLOCK FOR NEXT POUR

STONE BASE IS COMPACTED AND FLATTENED WITH RAKE AND BOARD

3" TO 4"

FOR BEST RESISTANCE TO WINTER FROST HEAVES FROM GROUND- LAY DOWN AT LEAST 4" OF CRUSHED STONE, GRAVEL OR CINDER

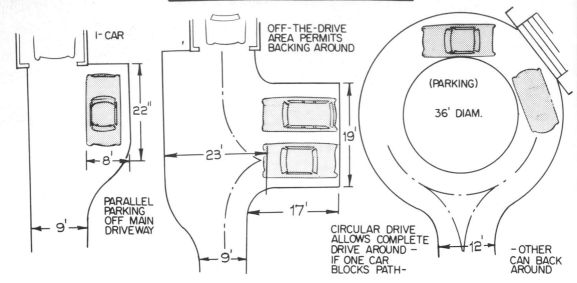

Think of a driveway as more than just for getting your car into the garage and you have the idea. This spacious drive gives room for turning, creates warm welcome.

FORM FOR CASTING CONCRETE BUMPERS

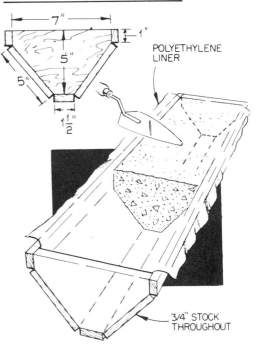

You can protect your plants, house, garage and lawn from wayward car tires by casting parking bumpers and placing them where they are needed most. How-to art, above.

Ideal do-it-yourself driveway contains a variety of slab sizes. It gives you a choice of how much work you want to do the day you pour it. Little time, small slab; large results.

Using colored or pebble concrete to set out traffic patterns you want cars to follow is both functional and attractive. Cast plain slabs first, to act as forms for the others.

shock absorber for frost-heave. If frost-heaving of your soil is a winter-spring problem, you'd best go for the subbase. A good place to find out is your local building department. They're not always right. But they are the best source of information on local conditions that you have. Many, however, tend toward overdesign on things like residential driveways.

The amount of driveway concrete should be figured the same as for a patio. Ready-packaged mixes are not economical for projects of driveway size. Use ready-mix or buy the materials and mix your own.

No steel reinforcing is needed in a well made concrete patio. Steel used in any reasonable amounts won't keep concrete from cracking. If you want to build in a margin of safety against cracking, add an inch to the driveway's thickness. It'll cost less than the steel, be easier and much more effective.

CONTROL JOINTS

Like all concrete, a driveway needs joints to control cracking from shrinkage of the material. Transverse control joints across the slab from edge to edge should be cut no more than 10 feet apart and to a depth of at least one-sixth of the slab—6-inch slab, 1-inch-deep joints. Dig?

A driveway wider than 13 feet needs a longitudinal control joint too. This may be cut down the drive's center or placed off center. No problem either way. Just so that no strip wider than about 10 feet is left.

Full-depth isolation joints should be provided wherever the driveway concrete meets other concrete walls or slabs. There should be one at the garage or carport. Another on both sides of the sidewalk. Still another isolation joint is needed at the point where the drive meets the street. Asphalt-impregnated isolation joint material is most convenient to form the joint.

Put a keyed construction joint at the end of each day's work, unless a good stopping joint is already formed.

SUGGESTED FINISHES

A broomed finish is especially effective on a driveway. Same with exposed aggregate. Both give a toothy surface texture that prevents tire-slip.

If, like many, you use your drive as a place to work on the car, you'll want a somewhat smoother finish than otherwise. This will make it more comfortable to work under the car and easier to roll under on a creeper. In this case I suggest using a magnesium float finish or a very fine broomed finish.

The photographs on these pages show some of the best current ideas in concrete driveways. Pick them over for ideas that you can use. A good driveway design idea will help make your house outstanding in the neighborhood.

A shell of adobe-like slump blocks creates an attractive planter-driveway light. A post of the same materials comes up through the center to hold light, space in front is planter.

DRIVEWAY LIGHTS MADE OF MASONRY

Here is a touch of grandeur you can add to your home
without a great deal of expense—and with little effort

HERE ARE two concrete-block entrance lights that you can build. Both are guaranteed to brighten your house's exterior.

Build them on concrete footings 8 inches thick and slightly larger than the outside of the project. Place the footing below the frost line or at least deep enough that the first course of blocks will start below grade.

The lights could well be made of bricks or stone rather than blocks. If you use bricks, be sure they are the kind that can take exposure at grade. Some can't.

The electrical conduit should be installed in the footing before you pour it. If the footing is deep, the conduit can enter above it, but still below ground. If you are a good home electrician, you can do the electrical wiring yourself using direct burial cable or wires. Otherwise, call in a professional to handle the hookups.

The slump block driveway light is actually a planter with the works for its light coming up through the center and supported by a square of masonry. Mount the lamp housing to its support using masonry anchors. The support should be brought up two courses above the planter for good light distribution. Access for bulb-changing and cleaning is provided by the lamp housing.

Fill your planter with a gravel drain layer, soil topping and plants.

GATE LIGHTS

The twin-post gate lights illustrate a good money-saving feature. Instead of purchased lamp fixtures, they use low-cost exterior light sockets mounted behind glass blocks. The glass blocks are laid just like masonry to form the gate posts. Access for bulb service is by removing the precast concrete post tops. A ridge underneath the tops keeps them from being slid off by accident.

Make the lights individually or as part of a fence, as shown in the photo, below.

Half-high concrete blocks were used to build this pair of gatepost lights. Their glass block fronts and rears cost much less than fixtures. Lamp service from removable top.

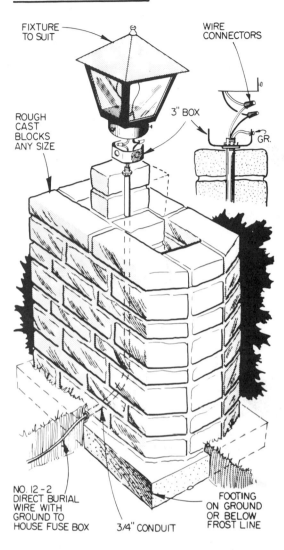

SLUMP BLOCK DRIVEWAY LIGHT

FIXTURE TO SUIT

WIRE CONNECTORS

ROUGH CAST BLOCKS ANY SIZE

3" BOX

GR.

NO. 12-2 DIRECT BURIAL WIRE WITH GROUND TO HOUSE FUSE BOX

3/4" CONDUIT

FOOTING ON GROUND OR BELOW FROST LINE

55

USING FABULOUS FERRO-CEMENT

You may well be first in your area to try this terrific new concrete medium

FERRO-CEMENT is an old medium that's had lots of promotion for use in building concrete boats—yes boats—but very little else. Yet ferro-cement is perfectly adapted to making many other projects, as long as their shape can be formed easily with wire mesh.

What is ferro-cement, you ask? It's super-reinforced thin-shell concrete made by troweling sand-mix concrete into layers of wire mesh. The mesh may be hardware cloth, chicken wire or metal lath.

Although a ferro-cement section is only perhaps ¼ inch thick, it's as strong as wood the same thickness. The beauty is that no forms are needed to make a ferro-cement project. The mesh serves as the form *and* the reinforcing.

Italian Professor Pier Luigi Nervi created and named the first ferro-cement in the late fifties. He used it to build folded-plate roofs on big buildings without forms. The medium was quickly picked up by amateur and professional boat-builders. Today there are hundreds of ferro-boats floating around the world. They're lighter and stronger than similar hulls in almost any other material.

You can use ferro-cement to build ferro-lawn-furniture, a ferro-flagpole, a ferro-house-sign, a ferro-fence or . . . well, use your imagination. The field is wide open.

Planter is typical of projects you can build of ferro-cement. Made 2′ in diameter, it takes advantage of the full mesh size. Color is integral concrete pigment, mixed well.

To make the planter, the wire reinforcement method was used. Wire was looped and twisted at the same time by feeding the end through the center of coil with each twist.

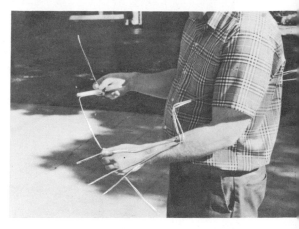

Four sides were made in the dogleg shape. The ends were left extra long to wrap around the chine, gunwale and base coils later on (see how it's done in next photo).

Two 2′ dia. coils of wire for the chine and gunwale and two 18″ coils for the base are wired to the side members by twisting the sides twice around with pliers, shown above.

The wire cage is placed on the mesh and rolled until the mesh is four layers thick. Author believes the cage might have been omitted, the wire mesh simply rolled up.

SHAPES IT MAKES

Any project made up of one or more of the following will make it in ferro-cement: flat surfaces, boxes, pyramids, curved surfaces, cylinders, cones, and parts of pyramids and cones. You can even form complex curved surfaces that are made up of straight lines, such as a hyperbolic paraboloid.

Think about it and lots of shapes will come to you. Just figure a way to build your birdbath or flowerpot out of them (make clay models) and you've got it.

Each figure is formed separately in layers of mesh. Then the separate figures required to make your complete ferro-project are wired together.

The best-working mesh we've found is what's called *aviary netting*. It's nothing more than fine-mesh chicken wire, having a ½-inch mesh. Hardware cloth is excellent too, but it costs quite a bit. If you use it, try ¼- or ½-in. mesh. Chicken wire, incidentally, can be made into various shapes by pinching the wire together.

When the desired number of layers is rolled, cut the mesh and wire down the flap with short lengths of mechanic's wire. Tuck wire ends in mesh so they won't poke out.

Best way, author found, to mortar the mesh was by hand, wearing vinyl-dipped gloves. While one hand "trowels" in the sand-mix concrete, the other smooths inside.

The thickness of your project should be governed by what stresses are to be put on it. A lawn chair obviously needs more strength than a birdbath. With proper use of shapes, the strength comes from the shape as well as the thickness of material. With ferro-cement shape is the best way to get strength. An eggshell's walls aren't thick, but the shape is strong.

Far as we know, no one has developed engineering data to help in setting thicknesses to use. Even in ferro-boat building, designers of new hulls have to guess at the section thicknesses required. You will have to do the same. Use lots of curves in your ferro-cement project. They create strength.

About the minimum thickness you can get your mortar to hold into is four thicknesses of aviary netting. Any less and the mix tends to push right on through or fall out. Greater thickness and greater strength is obtained with more layers of mesh.

FORMING

You have a choice of forming mesh with or without steel rods. The rods, if used, help hold the mesh in the desired shape.

They also add strength at the edges. Their drawback is one of cost and time to form.

We also tried using many strands of wire in place of rods. It was a great money-saver because we already had the wire in huge rolls.

For shapes than can be formed without rods or wire, I recommend eliminating it.

Getting the mesh into the shape you want is the only skilled part of the process. Feel free to use whatever system works best for you. We found cylinders and cones easy to make by rolling them out on the driveway and wiring to keep them from unrolling again. You may roll all your layers of mesh up together or roll until the shape builds to the numer of layers wanted.

Squares and rectangles were easy to form by bending the mesh over a straight-edge. Tough to form were compound curves, such as spheres. That's why we recommend making only simple figures. If you find a way to solve compound curve-forming, you've opened ferro-cement to building almost any shape. (Please let me know about it.) You could conceivably make a ferro-car. (Just let someone run into you in the parking lot.)

The partly mortared planter can be turned on its side for mortaring the bottom, if done gently to avoid dislodging mix from the mesh. It's best to do the bottom first, as shown.

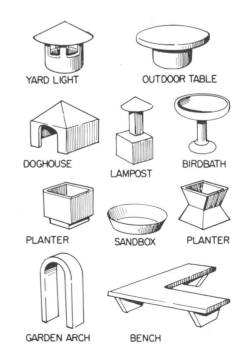

The edges of most ferro-cement projects need extra attention. Take pains to get them well filled with concrete mix. Your finger-thumb joint makes a perfect applicator.

EASY TO CONSTRUCT SHAPES

CYLINDERS   CONES

FOLDED PLATES   CURLED DECOR

FLAT PLATES   BOXES

PYRAMIDS   HYPERBOLIC PARABOLOID

COMBINATIONS THAT MAKE USEFUL ITEMS OUTDOORS

YARD LIGHT   OUTDOOR TABLE

DOGHOUSE   LAMPOST   BIRDBATH

PLANTER   SANDBOX   PLANTER

GARDEN ARCH   BENCH

59

Although this fence is made of masonry and precast concrete squares, it could easily have been built of ferro-cement. Ferro-cement can help you build many impressive projects.

ADDED BRACING

Before you finish the wire "cage," study it for stress points. A joint between surfaces where loads will be transferred from one surface to another is a potential failure spot. Strengthen by adding strips of mesh along the joints on both sides. Attach them with short lengths of mechanic's wire with the ends twisted together. The added mesh will hold more concrete at that point and thus make a stronger joint.

NOW THE CEMENT

Once your mesh cage is made and its parts wired together, go over it for spots where the mesh layers part and need to be wired together. Insert U-shaped wires from one side and twist them on the other with pliers. Be sure to lay the wire ends away in the mesh layers to keep them from grouging your hands during mortaring.

Sand-mix concrete for making ferro-cement is best mixed from ready-packaged mix. It should contain no stones, just cement and sand. A consistency that makes good concrete is right for this, too.

Masonry mortar would work for ferrocement, but it doesn't get as strong as sand-mix, and so is not recommended.

The easiest way to get the concrete into the mesh is with your hands. The alkali in concrete cement is hard on skin. Wear vinyl- or rubber-dipped gloves. Pick up a gob of concrete and spread it in with one hand while the other backs up the opposite side of the mesh. Try to get all the mesh fully covered with no mesh show-through.

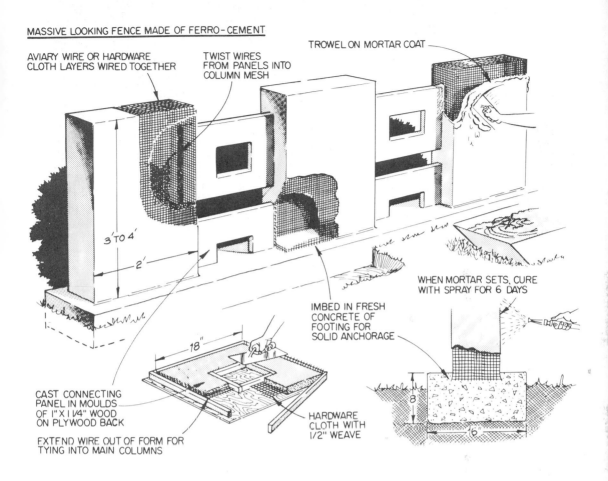

MASSIVE LOOKING FENCE MADE OF FERRO-CEMENT

TROWEL ON MORTAR COAT

AVIARY WIRE OR HARDWARE CLOTH LAYERS WIRED TOGETHER

TWIST WIRES FROM PANELS INTO COLUMN MESH

3'TO 4'

2'

WHEN MORTAR SETS, CURE WITH SPRAY FOR 6 DAYS

IMBED IN FRESH CONCRETE OF FOOTING FOR SOLID ANCHORAGE

18"

CAST CONNECTING PANEL IN MOULDS OF 1" X 1 1/4" WOOD ON PLYWOOD BACK

EXTEND WIRE OUT OF FORM FOR TYING INTO MAIN COLUMNS

HARDWARE CLOTH WITH 1/2" WEAVE

8"

6"

TOPPING LAYER

If your ferro-project ends up with any show-through of mesh, you can cover it later with a sand-mix topping. First dampen the surface. Then brush on a cement-water grout mixed to ice cream consistency. Before that dries white, spread on more sand-mix concrete made the same as the initial mix. The two layers will bond together into one.

Because a ferro-cement project is made with portland cement, it needs curing. Cover it with a sheet of polyethylene and leave it for a week. Afterward, it is ready to use. It will gain strength—as with any concrete—for 20 years and more. Treat it gently for the first month or so, and you'll be amazed at its strength afterward.

We load-tested a young ferro-tabletop. It held far more people than reasonably could have been expected. It finally cracked under the weight of six people but the crack held tightly. *That's* reinforcing!

One thing about building a ferro-project: you are a pioneer. Whatever you do probably hasn't been done before. Use my designs in any way you like, or adapt them as you wish. No one holds any ferro-patents that I know of. You should find no roadblocks to making and even selling your own designs in ferro-cement. Go to it.

Editor's note: I used this technique five years ago in making a piece of sculpture representing buildings in NYC and didn't know what to call it. It stands up fine. I only hope NYC does as well.

Sakrete, Inc. photos
A screen wall easily built of concrete grille block units gives privacy to a side patio. Build its concrete footing in a trench dug down below the frost line to prevent cracking lines.

BUILDING GARDEN WALLS AND FENCES

Besides looking expensive, a concrete or masonry wall needs no painting or care

A WALL OR FENCE can do quite a few things for your home. Depending on how you build it, you can establish property lines or keep people out. You can keep passers-by from seeing in, yet permit breezes to flow through. A wall can serve as a background for your flowerbed, provide something for plants to climb, protect from the sun and wind. When you figure out what you want the fence or wall to do, then you can design and build it.

The masonry rule-of-thumb is that a wall can be 18 times as high as it is thick. Most fences and garden walls are 6 to 8 inches thick. Thus, they could be 9 to 12 feet high. These are maximum of course. It's better not to go nearly that high. The problem is one of safety. If your wall is too high and not properly designed and built, it could blow over, damaging property and possibly injuring someone. So, as with most big projects you build, check your local code before you put up a masonry wall or fence.

In most communities you can build a wall up to 4 feet high without running into code problems. Some codes restrict walls higher than that, requiring that they be engineered, that is designed by a licensed professional engineer in your state.

Dig holes and set 4x4 redwood end posts. Shovel Sakrete sand-mix concrete in around the plumbed posts. A course of standard concrete blocks brings wall up to grade.

Lay screen blocks on top of the standard blocks beginning at grade. First step is to place a length of ladder-type block or brick reinforcement for a 4-in. below grade wall.

Spread workable Sakrete mortar mix on top of the reinforcement, about two blocks ahead. Watch the pattern as you lay each unit. One wrongie spoils the whole job.

Butter mortar onto the contacting edge of a block, being as careful as you can to keep it from getting onto the face of the block. Care here saves cleaning up time later on.

Set the block into position in the mortar bed and close to the previously laid block in the same course. Having a good-working mortar mix will help the job along immensely.

Tap the block into final position with the handle of your trowel. If the mortar is at the right consistency, it will go without hard tapping, yet not too easily. Make it work.

Pick up excess mortar oozing from the joints with your trowel used in a slicing/scraping motion. Sling the mortar back onto mortarboard for reworking and reuse.

After a time, tool the mortar joints with a rounded or V-shaped tool. Tooling not only makes the joints look better, it compacts the mortar for a stronger, weatherproof joint.

While you're at it, ask about required set-backs from property lines. Sometimes a fence may be placed on the line, other times it must be entirely on your side of the line—unless your neighbor agrees otherwise. Sometimes a fence cannot be built within a certain distance of the property line. In this case if you build one, you'll give up some of your lot area. Set-backs from alleys, sidewalks and buildings are often required.

SCREEN WALL

Walls are easiest to build of masonry. Casting a concrete wall is hard work because of the required forming. A masonry wall may be solid. Or it may be built with grille blocks, or of bricks laid with lots of openings in the pattern to become a screen wall. Screen walls let air through while giving a degree of privacy. Certain designs let you see out somewhat if you're close to

Reset the string line for a new course, using the end posts to hold it. Measure up from the last course the height of one block plus the thickness of one mortar joint, as above.

Before painting and landscaping, the finished screen wall looks raw. Note the course of standard concrete blocks used to bring the wall up to grade inexpensively.

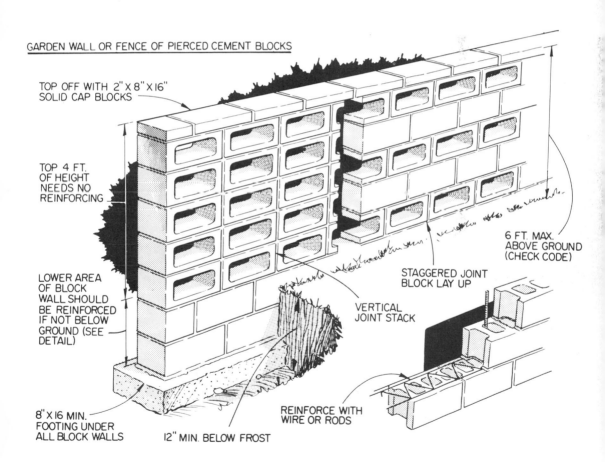

GARDEN WALL OR FENCE OF PIERCED CEMENT BLOCKS

TOP OFF WITH 2" X 8" X 16" SOLID CAP BLOCKS

TOP 4 FT. OF HEIGHT NEEDS NO REINFORCING

LOWER AREA OF BLOCK WALL SHOULD BE REINFORCED IF NOT BELOW GROUND (SEE DETAIL)

8" X 16 MIN. FOOTING UNDER ALL BLOCK WALLS

12" MIN. BELOW FROST

REINFORCE WITH WIRE OR RODS

VERTICAL JOINT STACK

STAGGERED JOINT BLOCK LAY UP

6 FT. MAX. ABOVE GROUND (CHECK CODE)

Another use of concrete grille blocks creates a wall high enough for both privacy and security. Concrete masonary pilasters are spaced every eight feet along this wall for strength.

Concrete masonry wall of half-high blocks forms the fence backdrop for planting in a desert backyard. Stacked bond pattern permits the use of vertical steel reinforcement.

them without giving distant passers a view in. They are especially good in a garden.

Every wall, low or high, should be built on a concrete footing cast in the ground below frost depth. Make the footing 8 inches thick and twice as wide as the wall is thick. The wall should, in most cases, be 8 inches thick or more. Build up from the footing to grade with plain concrete blocks. They lay up quickly and don't cost a great deal. From grade, go up with the units you want for the wall.

Lacking restrictive codes, your wall can safely be 6 feet tall. The top 4 feet of it need not be reinforced. The lower portion, however, needs reinforcing with ¼-inch steel rods placed in the horizontal joints. Those can be laid in every course of block (or every 8 inches vertically). Vertical reinforcing should be of ½-inch rods cast into the footing every 4 feet and bent upward. Fill the cores around the rods with concrete. Walls without units having cores can be reinforced by laying a stacked bond so that ¼-inch diameter reinforcement may be placed in the continuous vertical joints.

A course of grille blocks atop a fence wall helps to take away from the prison wall appearance. Here the grille blocks are the same size as the solid units for easy fit.

A 16'-long screen wall shelters the front entry from street and drive and backstops the flowerbed. Slump block pilasters are necessary to brace it against tipping over.

STRENGTH WITHOUT STEEL

Unreinforced walls taller than 4 feet can be made by building *pilasters* in the wall. Pilasters are pillars or columns laid up with the wall and bonded to it. They should be twice as wide as it is. There should be a pilaster along the wall at least every 18 times the width of the wall. For example, if the wall is 8 inches wide a pilaster must be built every 12 feet or less.

Pilasters or reinforcing are necessary to brace a tall wall against being blown over in a high wind. You can get the same bracing effect by making the wall serpentine or zigzag. Such a wall could probably be built without any pilasters or reinforcing steel. Check your code.

Free-standing nonserpentine walls, that is those without bends or corners to brace them against blow-over, also should extend at least 18 inches below the ground. That way the ground provides a bracing.

CAPPING

A wall may be left uncapped. Or a course of solid cap block or row of sidewise bricks may be laid along the top of it. In addition, cut stone can be used as a capping. You can also form and cast a

Here the grille blocks are different sizes from the solid units used to build a wall. Note that a final course of solid units was used to cap wall for unified appearance.

Some structural concrete blocks make excellent screen walls. Here 16" square chimney blocks were laid on their sides to make an 8" thick wall. Entire wall was painted.
Robert Cleveland

High-type concrete slump blocks were combined with standard concrete blocks in interlocking courses. This trick saves both money and time in the building of the wall.

Slump block columns laid on concrete footings hold drilled 2x4 rails. Lengths of closet pole stock inserted through the rails are used to create an open picket-like fence.

Bricks or concrete blocks without openings in them can be laid in unusual patterns to build a screen wall. Reinforced concrete top beam is needed for strength of wall.

A no-maintenance rail fence is made with slump block columns and precast concrete rails. When you cast the rails, put a length of reinforcing rod down centers of blocks.

Wrought iron gate fastens into holes bored in the blocks. Masonry anchors and lag bolts were used to fasten the frame to the blocks. Gate and wall were painted for finished job.

reinforced concrete bond beam along the top of a long wall to strengthen it. The block capping course is easiest. Walls made of hollow-core units need some kind of capping to keep water out of the course where it could freeze and break up the wall from within. A cap is optional otherwise.

FENCES

If security and privacy are not the purposes of a wall on your property line, you can build it more as a fence than a high wall. An excellent fence that saves much work is one built of masonry columns. These hold wood or precast concrete rails between them. They go up quickly and look great. The columns should be placed on concrete footings below frost and capped so that water runs off. Holes are left in the columns for setting the rails. Mortar fills any excess spaces in the masonry around the rails.

Naturally, if wooden rails are used, maintenance will be needed to keep them looking good. The longest-lasting finish is stain. Avoid clear finishes. They don't last outdoors worth a darn.

When you wall in your property to keep others out, remember that you also wall yourself in. This is true of your view too. To restrict the view in, you must also restrict your own view out of the house.

GATES

You'll need one or more gates to let family members in and out. These can be locked, or wired to an alarm, as you wish, to control their use. Most practically for your own use, the gates are simply latched to keep small children and animals from going through.

Wrought iron makes excellent appearing gates with a masonry wall or fence. Wood doesn't seem to match up. The advantage of a wood gate is that you can make it yourself. This can be a big benefit in saving you money. Wrought iron work doesn't come cheap. On the other hand, a wrought iron gate needs less maintenance —painting—than a wooden one. Take your pick.

Another gate material to consider is ferro-cement. We can see no reason why an excellent gate can't be made using this relatively untried medium.

The gate's appearance would match that of the wall. See the chapter on ferrocement for information on designing and working with ferro-cement. Gate hardware could be standard, probably, with the hinges and latch incorporated structurally into the ferro-cement by both wiring and bonding. Bet you'd be the first in the world to have a ferro-cement gate. It'd be a terrific conversation starter. Think about it, and start designing.

69

Robert Cleveland

Using metal molds you can cast concrete garden objects that would be tough to form in any other way. The surface texture of the planter-urn above was made with a stony, dry mix.

CASTING GARDEN ART IN METAL MOLDS

Making your own or making extra money, casting garden art is rewarding

CASTING IN METAL molds could be a whole book in itself. Here we can only introduce the subject. Instructions that come with your metal forms will tell how to use them.

Most forms are aluminum. They're made for such objects as urns, birdbaths, planters, statues, pedestals, benches, plaques, tables, animals, lanterns, fountains, lampposts and columns. Get the forms from firms such as Concrete Machinery Co., Inc., Drawer 99, Hickory, N.C. 28601.

Two different methods of casting are used. The *pouring method* uses a soupy concrete mix that's poured into the form and allowed to settle. More is poured in to keep the form full until the settling stops. The pouring method is best for closed-up top forms for making slender and intricate objects.

The hand-packed method uses a very dry mix that is tamped into the mold in layers. It can be used only with open-top forms that have access for tamping. A poured object takes days to harden, while the forms for a hand-tamped project can be stripped the same day as casting.

Concrete Machinery Co. photos

Getting ready for casting a hand-packed object, the dry sand-mix concrete (left) and hand tamper (center) are placed next to the aluminum mold on the work table surface.

Part of the form for a concrete flower urn is removed after the urn has hardened sufficiently. Other parts of the form are simply unbolted and tapped loose with a mallet.

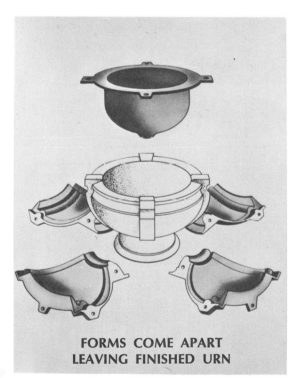

**FORMS COME APART
LEAVING FINISHED URN**

Concrete Machinery Co.

Concrete bench with intricate scrollwork design is easily made in two forms, one for the top and another for the legs. The parts are held together with dowels and epoxy.

Japanese Ming lantern is cast in five forms and the pieces assembled with epoxy cement. Such products sell for many times their cost to make. It's a very good business.

71

Slump block masonry planter built against one wall of the house becomes a part of the home and its landscaping. No rear wall for the planter is needed with a masonry house.

PLANTERS FROM THE GROUND UP

You can do wonders around the house with large and small masonry planters

A MASONRY PLANTER is simply a permanent, nonmovable flowerpot for landscaping your house and grounds. A planter built in conjunction with a screen wall of grille blocks makes an excellent combination. Use it at the front, back or side of your house to protect an entry or add privacy to a patio.

Because a planter is permanent, each masonry planter needs a footing. Put it below frost where it won't heave and crack the planter.

If the planter is small, the footing may be a solid slab 8 inches thick. If the planter is large, you can save on footing concrete by using 16-inch-wide footings under the walls of the planter. In either case there's no need to form for the footings. Just dig the excavation for them to the right dimensions and let it serve as the form. Of course, if the soil is loose, the hole will not hold its shape and you'll need forms to avoid wasting footing concrete. The forms can be made of 2-inch lumber.

Planters out away from the house need walls on all sides to hold in the soil. One built against a masonry house can use the house for one wall. A planter next to a frame house needs a wall at the back too. And, to keep rain and sprinkling water from running in behind the planter and rotting out the house wall, a flashing is needed across top of the planter as protection. The house siding should be cut out and the flashing slipped up under it. One lip of the metal flashing extends out over the top course of masonry when you lay it. Calking between the masonry and flashing keeps all water out.

To be sure no water gets in at the planter's sides between it and the house wall, this joint should be calked, better yet, flashed and calked.

To protect the plantings you put in it,

your planter must have good drainage. It cannot be an enclosed tub. If cast on a slab, the planter should have lots of drain openings at the bottom. An easy way to make these is by leaving the end joints between units open with no mortar in them. Then put a layer of gravel or crushed stone in the bottom of the finished planter to conduct water to the drains. No drains should be placed where they would run water against the house.

You may want to put a layer of plastic screening over the gravel before filling your planter with soil. The screen keeps soil from working down to fill up the spaces in the gravel.

FROST-PROTECTION

The biggest destroyer of outdoor masonry planters is frost. No matter how strong the wall, it cannot resist the force of freezing soil. If you live in a freeze-thaw climate, you must protect your planter during winter. One way is to remove the plants and soil in the off-season and leave the planter empty. Make sure the drains stay open so that it doesn't fill up with rain water or melted snow.

A better way is to design your planter with in-built freeze-protection. Place a sheet of 1-inch-thick styrene foam or polyurethane foam around the inside of the planter snug against the masonry. Then place the soil in. Then foam will harmlessly take up expansion from freezing. No wall-cracking.

Another, more costly method is to build a sheet metal container to fit the inside of your planter. Make it slightly smaller to allow for frost expansion.

The second most common damage comes to masonry planters through being hit by automobiles. Don't build a planter close enough to the garage or driveway that a car can hit it. This is one of those

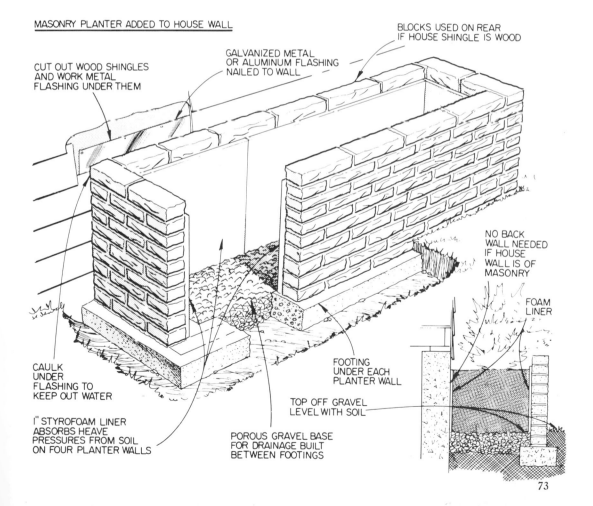

MASONRY PLANTER ADDED TO HOUSE WALL

BLOCKS USED ON REAR IF HOUSE SHINGLE IS WOOD

GALVANIZED METAL OR ALUMINUM FLASHING NAILED TO WALL

CUT OUT WOOD SHINGLES AND WORK METAL FLASHING UNDER THEM

NO BACK WALL NEEDED IF HOUSE WALL IS OF MASONRY

FOAM LINER

CAULK UNDER FLASHING TO KEEP OUT WATER

FOOTING UNDER EACH PLANTER WALL

1" STYROFOAM LINER ABSORBS HEAVE PRESSURES FROM SOIL ON FOUR PLANTER WALLS

TOP OFF GRAVEL LEVEL WITH SOIL

POROUS GRAVEL BASE FOR DRAINAGE BUILT BETWEEN FOOTINGS

A two-level garden hobbyist's planter is created with formal looking stacked block pattern. Long, slim vertical units make the large circular planter for around tree base.

Modest planter made to hold potted plants is a size that requires no gardening proficiency of its owner. With the plants in pots, no back wall is needed on the planter.

occurrences that if it *can* happen, it *will*. At least put precast concrete parking bumpers to keep cars from pulling too close to your planters. It will save both your planters and cars.

MATERIALS TO USE

A masonry planter may be made of bricks, concrete blocks or stone. There's no limit on size of units. The larger they are, the fewer there are to lay. You shouldn't, however, use porous bricks to build a planter. They perform poorly in contact with soil. They can't take freezing and thawing. Use hard-weathering bricks instead.

Reinforcement is not needed unless there is some special reason for it. The steel is not sufficient to prevent the walls

from cracking under freeze pressures. As you can imagine, grille blocks aren't very practical units for planter construction unless you back them up with solid masonry. Then they'd give a decorative effect. Better to save grille units for walls where they have a more functional use.

CONCRETE PLANTER

Masonry is not the only material suitable for making planters. Not as easy but still possible, building a concrete planter can be fun. The combination concrete planter-entry shown in the illustrations is made in three steps. This builds a planter, plus an access walk, plus steps up to an entry slab just outside the door. Using the same techniques you can make many major concrete planter projects around your

Here's a brick planter you can make while waiting for the ball game to come on. You can build it three bricks square or vary the dimensions by full or half-bricks to suit.

Fieldstone planter has a seat at one end. In laying the stones, an attempt was made to get a flat ledge along the front wall to make weeding of the planter more comfortable.

Curving walls can be made from straight units. Here a planter has been stuck in an unused corner of a parking area. Units were selected to match the wall and to tie it in.

If you're going to have a window box, HAVE a window box. The unusual texture was made with recessed-edge concrete blocks. A woodsy window is created, even in town.

house. Just try not to overdo a good thing.

The first job is to prepare the excavation. Make it with 4-inch-thick slabs in mind. Decide on what your finished grades are to be at every level and excavate 4 inches below them. Try not to dig out too much, because it's better to place concrete on undisturbed ground than on fill, even though the fill is well tamped to compact it.

Step one is to form the planters, as shown in the drawing (top). Both of them reach down below the slabs to the lowest grade and rest on it.

Since the planters are about 16 inches high, plywood is used to form them. Stakes hold the lower portion of the forms and 2x2-inch form wales are nailed on to hold the upper portion. Concrete pressures are outward; the wales are needed to resist. If

they are not stiff enough for long spans of form, you can nail cross strips of 1x2-inch lumber over the tops of the forms to hold them in. These get in the way of finishing the top, so avoid them if possible.

The opening in the large planter is boxed out with a form of plywood too. Oil all forms for easy removal.

SAVE ON CONCRETE

Pour your concrete and compact it along the forms with a shovel. Tapping the forms with a hammer helps to remove air pockets from the formed surface.

Finish the top surface as desired.

Concrete need not be 16 inches thick and so you can save on mix, if you wish, by placing large boulders or broken concrete into the form. Keep it 4 inches back

Good use of materials, this concrete block sidewall planter is made of cap blocks. For a rustic effect mortar joints were not trimmed. Blocks set on end make back wall.

Small front yard? Fill it with planters. The problem of changing elevations is solved by elevating the planters. Be sure you don't cover up hose outlet in your enthusiasm.

Robert Cleveland

Impressive concrete planter-entry can be added to any house by thinking big. Massive-seeming planter with its thick walls is cast first, then steps, then lower slab is cast last.

from the forms and the top of the planter.

After the planter has cured, remove all forms and form stakes.

Now cast the steps and upper-level entry slab as a unit. The steps make this the hardest part of the project. If you have friends willing to help, call on them.

Part of the planter and curb wall serve as forms. The other forms can be made of plywood and 2-inch and 1-inch lumber, well staked in position; 1x4-inch boards placed across the step forms brace them.

Pour the concrete and remove the stakes that are buried in it after strikeoff. Chunk fresh concrete down into the void left after stake removal and work it in with the rest of the step concrete.

If you want trowel-finished stair fronts, you'll have to pull off the step forms when the concrete has stiffened sufficiently. We prefer leaving a formed finish on steps. It beats pulling the forms too soon and having half the step slump away. (The patch never seems to look right.)

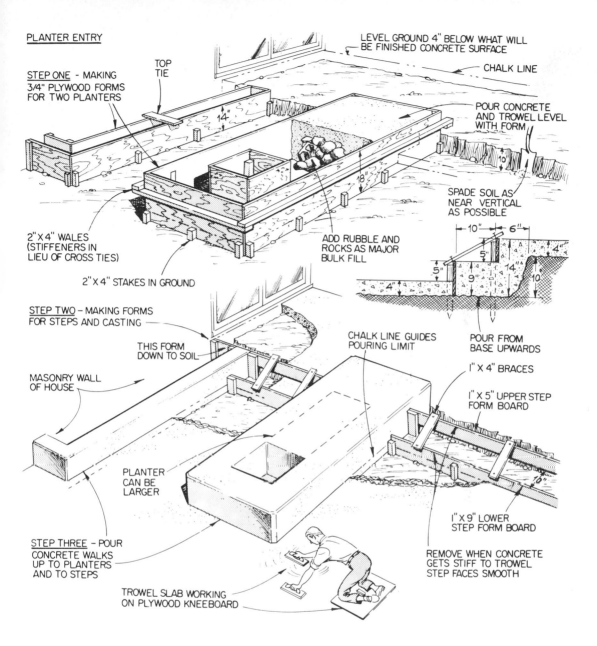

PLANTER ENTRY

STEP ONE - MAKING 3/4" PLYWOOD FORMS FOR TWO PLANTERS

TOP TIE

LEVEL GROUND 4" BELOW WHAT WILL BE FINISHED CONCRETE SURFACE

CHALK LINE

14"

POUR CONCRETE AND TROWEL LEVEL WITH FORM

10"

18"

SPADE SOIL AS NEAR VERTICAL AS POSSIBLE

2" X 4" WALES (STIFFENERS IN LIEU OF CROSS TIES)

ADD RUBBLE AND ROCKS AS MAJOR BULK FILL

2" X 4" STAKES IN GROUND

10"   6"

5"

4"

5"

9"

14"

10"

4"

STEP TWO – MAKING FORMS FOR STEPS AND CASTING

THIS FORM DOWN TO SOIL

CHALK LINE GUIDES POURING LIMIT

POUR FROM BASE UPWARDS

MASONRY WALL OF HOUSE

1" X 4" BRACES

1" X 5" UPPER STEP FORM BOARD

PLANTER CAN BE LARGER

1" X 9" LOWER STEP FORM BOARD

STEP THREE – POUR CONCRETE WALKS UP TO PLANTERS AND TO STEPS

REMOVE WHEN CONCRETE GETS STIFF TO TROWEL STEP FACES SMOOTH

TROWEL SLAB WORKING ON PLYWOOD KNEEBOARD

LAST IS EASIEST

The last step in building your planter-entry is placing the lower-level slab. Since it is a slab and nothing more, handle it like any other slab. If it is a large slab where you cannot reach all portions from outside, you'll have to float and finish on knee-boards. You can make them of a 16-inch-square piece of ¼-inch plywood or hardboard with 1x2-inch pieces nailed on as handles. Put your knees on one board, your toes on the other. Move the kneeboards one at a time, always finishing over the marks left by them.

Since you don't need any forms along the surfaces of your planter and curb, snap chalklines there to guide you in leveling. Be sure the slab slopes to drain.

Stand back and admire your finished planter. You'll have completed a major home improvement and you'll be surprised at how easy it was. Especially if you took it in easy stages.

BIRDBATH IN THREE PIECES

Two forms are all you need to precast this neat three-piece concrete birdbath

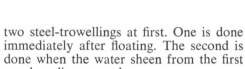

HERE'S AN EASILY stored concrete birdbath that comes apart into three flat slabs of concrete. During the off-season they can be kept almost anywhere. Set up in your yard and filled with water, the birdbath provides a 2-inch-deep swimming pool for feathered peepers.

Make the forms of ½-inch plywood and 1-inch lumber. The way the forms are set up, one side of each base piece and the bottom of the top piece gets a trowel finish. The rest gets a form finish. For this reason, you'll want your trowel finish to match the form finish as closely as possible. Try

two steel-trowellings at first. One is done immediately after floating. The second is done when the water sheen from the first one has disappeared.

A form for the well in the top is made of ¼-inch plywood or hardboard cut into triangles that roughly form a pyramid with a 2-inch rise (see photo). Nail the pyramid to the form with finishing nails that won't leave marks on the finished casting.

It takes about $3 worth of concrete to cast the bath. If you assemble the form edges with screws, the form can be used over and over.

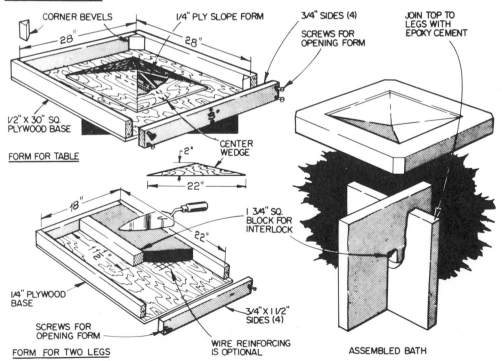

THREE PIECE BIRDBATH

CORNER BEVELS — 28" — 1/4" PLY SLOPE FORM — 28" — 3/4" SIDES (4) — SCREWS FOR OPENING FORM — JOIN TOP TO LEGS WITH EPOXY CEMENT

1/2" X 30" SQ. PLYWOOD BASE

FORM FOR TABLE

CENTER WEDGE — 2" — 22"

18" — 1 1/2" — 22" — 1 3/4" SQ. BLOCK FOR INTERLOCK

1/4" PLYWOOD BASE

SCREWS FOR OPENING FORM — 3/4" X 1 1/2" SIDES (4)

FORM FOR TWO LEGS — WIRE REINFORCING IS OPTIONAL — ASSEMBLED BATH

FASTENING WITH EPOXY

Epoxy may be the best way to glue the parts of your concrete project together

ANYONE WHO WORKS with concrete and masonry sooner or later will find uses for the epoxy resins, those two-part, mix-together glues that bond to almost anything. The epoxies work very well on masonry and concrete. They can be used for patching, surfacing, filling or fastening things to walls and slabs. Epoxies ordinarily set faster than concrete and bond with greater strength than the original masonry surface. Unlike glue, no clamps are needed. However, the surfaces must somehow be held in position until they are bonded.

A number of brands of epoxies are sold by hardware and building supply dealers. They are made by Smooth-On, Woodhill, Duro, Linatex, Devcon and others. They're available in tubes and cans up to quarts and gallons. Some are designed for patching, some for adhering. Some will adhere fresh concrete to hardened concrete. Most have a pot life of less than an hour, so must be mixed only as needed for imme-

diate use. Toluol, denatured alcohol, methyl ethyl ketone (MEK) and acetone are useful for cleaning hands and tools. This should be done promptly after use because soon as the epoxy cures and hardens, it isn't practical to remove it.

BOND ALMOST ANYTHING

Epoxies will bond almost anything to concrete and masonry except rubber, nickel, tin, zinc, polyethylene and *Pyrex* glass. (Regular glass may be bonded.) They work especially well on steel, aluminum, wood and ceramics. They're highly resistant to water, most acids, most solvents, boiling and freezing.

For filling holes and cracks, epoxies may be extended for economy by mixing them with a filler. This may be powdered clay, silica (white) sand, chopped glass fibers or dry portland cement. Filler can be added until a consistency for puttying is reached.

EPOXY USED TO JOIN PARTS OF A BIRDBATH

All epoxies come as two-part mixes. To use, hardener is mixed with resin in proper proportions. Then the epoxy must be used up within its pot life of minutes, sometimes hours.

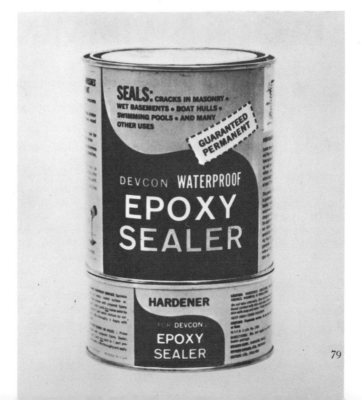

You can build a concrete block swimming pool in spare-time that will fill all of your family's leisure time from then on with the joys of backyard swimming. It's also highly decorative.

EASY-TO-MAKE BLOCK SWIMMING POOL

Here's an in-the-ground swimming pool that's
easy enough for almost any handyman to build

FIRST LET'S straighten out one thing. This swimming pool is easy to build, sure. But we mean easy *compared to other in-ground swimming pools*. You'll cuss me for calling it easy compared with other projects described in this book. Building any pool is, of course, a lot more work than making a sidewalk or driveway. However, it is worth a lot more when finished.

That settled, we can get on with the how-to-build part. The design and method of construction were developed by the Cement and Concrete Association, of Britain. The pool's design is strong, yet simple enough that most can do it. It may be made any size. Our instructions don't limit you. A 15x24-foot pool is recommended. With a level concrete floor, the water depth will be about four feet. If you want greater depth for diving, you'll have to build a well in the floor at the diving end. This adds to the work, but also adds to the pool's usefulness for adults and older children. We recommend it.

The reinforced concrete floor is surrounded by two tiers of 4-inch-thick concrete block. The walls act as forms for 4 inches of concrete placed between them.

If you do all the work yourself, except digging, you should be able to come out at about $1000 for the whole job, including a spacious concrete block deck. The necessary pool filter and fence is extra. It's easy to spend more than this on a large plastic-lined above-ground pool. And one of those doesn't add a bit to your property value.

TOOLS

Most of the tools needed, you already have. Things like spade, shovel, wheelbarrow, pail, hose, hammer. The only others required are a 10-foot straightedge—which you can make out of wood—a 50-foot tape measure, a 3- or 4-foot level, a carpenter's square, standard masonry trowel, lines and blocks and cement-finishing float and trowel.

It will help to have a concrete mixer or small 5-gallon-pail-type concrete mixer. Either one is handy for mixing mortar, whether or not you use ready mix for the main concrete.

SITE SELECTION

The site for your project is no doubt limited to your backyard. Find out your local code regulations as to setback from the lot line and other requirements before you start. They can be very strict on pools. Ask your city or county building officials whether you must apply for a building permit on your pool. In most localities you must. If one is not absolutely required, avoid getting a building permit because your property tax is sure to reflect the value of the improvement from then on. The same is true for any concrete or masonry project built as a permanent addition to your house or grounds. A permit means that the tax assessor will be around to consider increasing your property valuation. And an increased property valuation results in higher taxes. Increased spending by local governments is making taxes high enough as it is.

The pool site should be open to the south, southeast and southwest to benefit most from sunshine. It should be screened from the other directions to keep out cool breezes. Trees near the pool are a nuisance. They drop leaves and other debris into the water, adding unnecessarily to your work keeping the pool clean. Growing tree roots also have a way of getting into pool drains, and under its floor to cause problems.

POOL DRAIN

Another thing to consider in site selection is below-ground obstructions. You can't put an inground pool where the sewer line, water main or buried electrical wires run without moving these utilities. That

TRANSFERING LEVEL FROM MASTER GRADE STAKE

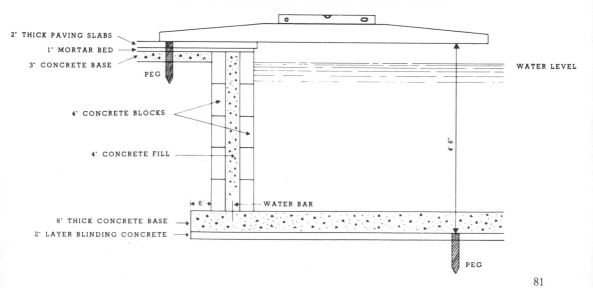

2" THICK PAVING SLABS
1" MORTAR BED
3" CONCRETE BASE
PEG
WATER LEVEL
4" CONCRETE BLOCKS
4" CONCRETE FILL
4' 6"
6" — WATER BAR
6" THICK CONCRETE BASE
2" LAYER BLINDING CONCRETE
PEG

Set a master grade stake just outside the pool excavation and transfer elevations from it to the floor. Straightedge, level, prop stick and tape measure are used here.

could be costly. Choose a site that's free of them. Keep the pool a safe distance from parts of a septic system too, for obvious reasons.

The pool location may be influenced by its method of draining. If it is to be drained by below-ground pipes, the bottom of the pool must be above the level that these can drain. You can get around this problem by using a motorized pump to drain the pool. Fasten a length of garden hose to each end of the pump and you can move the water pretty much where you want it. In this case, you'll want to build a sump in the pool bottom where you can place the suction hose from the pump.

The top level of the pool and its deck may have to correspond to the present level of the house, patio or backyard.

EXCAVATION

When you decide on the site, stake out the area for excavation. Add 1 foot 6 inches to the dimensions of the pool's floor slab. The additional size leaves room for the floor forms and for working around the outside wall when you build it. And if it

Place three rows of grade stakes along pool bottom with proper slope for drainage. Sight across three equal-height sticks placed on top to be sure stake rows are straight.

Start the blinding layer with a strip of fairly dry concrete. Make three foot-wide strips along the rows of grade stakes. Tamp and level the strips to exact elevation desired.

Now fill in the area between strips with dry concrete and tamp it to compact. Level with a straightedge, fill in low spots, tamp them to get the correctly sloped bottom surface.

rains while you're building, the extra space helps in pumping. For a 15x24-foot inside-dimension pool the excavation should measure 19 feet 6 inches by 28 feet 6 inches.

Drive wooden stakes showing the corners of the hole to be dug.

Excavation depth is measured from a master grade stake placed slightly outside the excavated area. Level measurements are taken from this stake (see drawing). It lets you keep close control of the excavation's bottom.

You should not try to dig the hole by hand. You'll find yourself earning something like 50 cents an hour . . . if you dig fast, that is. Hire a contractor with a backhoe to do the digging.

Stop the mechanical digging slightly above the final grade and do the last few inches by hand. This avoids disturbing unexcavated ground. Undisturbed earth makes a more stable subbase for the pool bottom than fill does. Do the final trimming by hand.

If it rains or the site is wet with ground water, you'll need a pump to bail out water. This same pump can be used later to empty your pool. You can rent a pump too, if you need one.

When working out your levels allow an inch or two for what's called a *blinding layer* over the pool floor. The blinding layer provides a properly graded work surface for laying the floor slab.

Dig the perimeter of the pool level to get the walls started level. The rest of the floor should slope about ½ inch to the center from both sides and the same amount to the deeper end, where the sump or drain should be located. It can slope steeply to a diving well at the deep end.

If the ground is wet, final trimming should be done just before placing the blinding layer. This keeps the foundation as undisturbed as possible.

POOL PIPING

If the pool is to have a bottom drain pipe, install it. Stuff newspaper in the pipe's end to keep stray concrete out.

Transfer the levels from the master grade stake to the pool bottom stakes using a level, straightedge and tape measure. Drive stakes in the floor to establish the grades for the top surface of the blinding layer. One row of stakes down the center

Place concrete for the floor slab in two layers on top of the blinding layer. Lay 6″x6″ steel mesh reinforcement on top of the first 4″ course, as shown in photo, above.

of the floor and others down each side is recommended.

BLINDING LAYER

Use a low-quality, cheap concrete mix for making the blinding layer. If you mix it yourself, proportion the concrete into a 1:3:4 mix: one part cement, three parts concrete sand and four parts ¾-inch gravel. If you order ready-mixed concrete, ask for a 4-bag mix.

The mix should be drier than you are normally used to seeing concrete. It should stand up in piles and have to be tamped to get it to do what you want. In ready-mix order 4-inch slump concrete. If you mix it yourself, watch your water use, keeping the mix dry.

The blinding layer should extend 6 inches beyond the floor slab all around to give support to the side forms. Lay strips of blinding concrete about 12 inches wide along the three rows of stakes. Compact the rows with a length of 4x4 that has a 6x6-inch square of plywood nailed to its end. Level the surface with a straightedge. It should end up even with the grade stakes. What you have done, is create three runners of concrete blinding that serve as guides for making the rest of the blinding layer. The stakes needn't be removed.

Now fill in the spaces between runners. Compact the concrete and level it across with a straightedge. Fill in any low spots

Use a length of 2"x4" or 2"x6" to compact and level the top course of floor concrete. Follow that with wood-floating and possibly with a steel-troweling for a smooth finish.

and tamp and level them for a good grade.

The whole blinding layer should be placed at once if possible.

CASTING THE FLOOR

The floor slab is too big a job to be placed all at once unless you have lots of help. Divide it into two or four slabs. Cast one at a time. With several eager helpers you can build the pool floor in two longitudinal halves.

Whatever, the floor should be reinforced with 6x6-inch steel mesh made with heavy enough wires to weigh about 5 pounds per square yard. Lap sheets of mesh reinforcement 12 inches between sheets. Stop reinforcing 2 inches inside the slab edges to prevent corrosion of the steel. The mesh should be continuous between all floor slabs. For this reason, you'll want to rip 2 inches off the 2x6 forms at these locations and to let the mesh fit between the strips (see photo). Hold the forms in position by nailing and with stakes made from 2-foot-long reinforcing bars driven through the blinding layer into the ground. If you want the best, most watertight floor, lay sheets of polyethylene over the blinding

layer before forming for the floor itself.

Floor concrete should be a quality mix. Use 1:2:3 concrete if you make it, or 6666 concrete if you order ready-mix. Maximum gravel size in either case should be ¾ inch, as for all the pool's concrete.

Place the lower 4 inches of concrete first. Level it by screeding. Then lay the steel mesh on top of it. Install the upper form strips over the mesh and place the top course of concrete. Give the surface a float finish as soon as it loses its water sheen. If you want a really smooth floor, follow this with a steel-troweling.

SUMP

If you want a sump in the floor, put in the box-shaped plywood form for it. The box is placed in a hole cut down through the blinding layer. Add a 1-inch wood lip around the top of the box to form a recess for the steel grating that will later cover the sump. The top of the lip should be set level with the top of the finished floor. Place floor slab around the box's sides.

The pool's design calls for a metal water bar around the edge of the floor. The seal is made from a 4-inch strip of galvanized steel. Position it 12 inches in from the floor edge so it will fall exactly between the two rows of blocks. Embed it 2 inches below the floor and leave the other 2 inches of width sticking up to engage the wall concrete placed later. Bend 90 degrees at corners. Overlap strips about 6 inches at joints. Stretch a stringline if it helps to get good alignment of the water bar.

Cure the finished floor slab to prevent drying out. I recommend a membrane spray cure, either sprayed or rolled on.

Forms can be removed about two days after concreting and the next slab may be case. Don't start the walls before the floor concrete gains sufficient strength. This takes about three days.

When you raise the mesh to remove the lower form, be careful not to damage the fresh concrete by rough treatment.

WALLS

The two tiers of pool walls are laid with 4x8x16-inch partition concrete blocks. For the 15x24-foot pool, you'll need about 675 blocks. The blocks should be kept dry until laid. Cover with polyethylene sheeting.

Clean the floor where the walls will meet it. Straighten the water seal if it got bent. Use form stakes to hold stringlines to mark the inside edge of the inner tier.

Lay the blocks as described in the chapter on laying masonry units, preferably using a good ready-packaged mortar mix.

Start laying blocks at the corners, going up several courses, then laying all the blocks in those courses between the corners. Bring up both outer and inner tiers at the same time.

Cut blocks with a brick chisel or block hammer as necessary to fit and to keep vertical joints in the block walls from coinciding with those in the same course below and in the tier opposite.

Place a wall tie every 3 feet along the rows of block. At their simplest these are thin strips of corrugated galvanized steel. If you want super strength, use wire block wall reinforcement. This is placed in the mortar joints too.

At every course in the corners put right-angle-bent reinforcing rods on top of the wall ties. Make these from 4-foot 6-inch lengths of ⅜-inch rebar. You may need to add extra wall ties at the corners to support

Make sump box form out of plywood. Coat outside of form with old crankcase oil so it will separate from concrete. Steel mesh is cut to go around the sump as shown here.

When form is stripped, a concrete-lined sump is left in the floor of the pool. A 1" lip on the sump's form creates a slight recess to hold the grating flush with the floor.

A galvanized steel water bar is wiggled down into the concrete all around where cavity between block tiers will go. Blocks placed temporarily to show bar position.

Drive stakes and stretch stringlines to mark the position of the inside edge of the inner tier of blocks. Check carefully for square with a carpenter's or a home-made square.

BONDING OF BLOCK COURSES

Suggested bonding in different courses, showing arrangement of blocks at corners.

the reinforcing. Judge it when you see it.

Don't carry the pool walls higher than six courses (48 inches). Even that height should be checked with your local building authorities. Higher walls need more reinforcement than is provided in this design.

SMOOTH WALLS

If you wish, you can get ground-face blocks with a colorful texture to use on the inside. Grinding adds about 10 cents a block to the cost. You can also use glazed-face units that look like tile when laid in a wall. Smooth-faced blocks present a better pool finish that will clean more easily. Use them, especially for the top course, if you can.

While you lay blocks, be careful to keep mortar drippings out of the wall cavity between tiers. You can rig a strip of wood placed over the cavity on the ties to catch drippings. When you finish laying the course above it, simply lift the strip out with the wires and dump the drippings.

Let the wall set up for seven days.

FILLING THE CAVITY

Much of the pool wall's strength comes from the concrete placed in the cavity between blocks. Use the same concrete mix as for the pool floor, but you'll probably want to mix your own, as placing goes so slowly it isn't a practical ready-mix job. Get a little extra water in the concrete for the lower 3 inches of cavity filling. This will help it to be worked tightly around the water bar. The rest of the concrete should be of good stiff consistency.

Fill the cavity in 12-inch lifts, working clear around with one lift before starting the next. A flat shovel works best for concrete placement. Use a steel rod, a pipe or a length of 2x2 wood to compact the cavity.

OVERFLOW

You can buy special masonry units that build into a pool gutter to catch overflow. Let the cavity concrete harden for two days before setting the overflow units in mortar across one end of the wall. Be sure to get the overflow channel perfectly level. If you don't, the water in the pool will show up the difference in elevation.

The open texture of concrete blocks should be given a seal of thick cement-water paste. Dampen the blocks and brush it on with a hand brush. Finally the blocks should be painted on the inside with a good pool paint (see the chapter on coloring concrete).

Instead you can make your own pool paint by adding a concrete coloring pigment to your grout application.

The weekend after the cavity concrete

has been placed you can backfill around the outside of the pool with excavated earth. Place the fill in 6-inch lifts and ram it hard with your tamper before placing the next lift. If you don't get good compaction during backfilling, the fill is sure to settle later. Then you'll have to remove your pool deck and refill under it. It's easier to do a thorough job now. Rent a power tamper if you wish. It will save lots of work.

POOL DECK

No swimming pool is fully enjoyable without a deck around it. One of the lowest-cost decks is made with paving blocks. Cast the blocks yourself or buy them ready-made.

The deck area will have to be excavated to allow for the paving blocks, plus the base for them. The 3-inch cast concrete

The outer tier of blocks is laid with a 4″ space between it and the inner tier. The two tiers are held together with wall ties placed in each course 3 feet apart, as shown below.

Spread mortar and lay the blocks, building both tiers together. Lay up the corners several courses high, then lay the blocks between the corners to complete all courses.

Like many other pool-building tasks, block-laying can be a family project. It goes best this way. Here smooth-faced blocks being laid in top course of almost finished wall.

After the last course of blocks has been laid, the cavity concrete is shoveled in and compacted with a tamper. Top course on the end wall is overflow blocks, shown below.

Pool will be more attractive and usable if you grout the inside to fill block cores. Adding color to grout before brushing it on creates your own functional cement paint.

Specially shaped blocks are laid for a pool gutter at one end. Gutter should be piped to drain overflow water to an out-fall or a dry well. Lay these blocks in mortar too.

Pool edging of specially shaped, textured units gives your pool a professional look. Lay the edging in mortar like bricks, getting perfectly level and straight all around.

Here paving blocks are being tamped into a 1″ mortar bed on top of a 3″ layer of deck concrete. If joints are to show, finish them ¼″ below the surface or use colored mortar.

base shown in the cross section drawing and photos makes the best base. You can also make a pool deck with sand, as shown in the chapter on paving a patio with bricks. Instead of paving blocks, the whole deck may be made of cast concrete or bricks. What you are doing is creating a patio around your pool.

Whatever, the deck should slope ⅛ inch per foot away from the pool edging. This drains surface water away rather than into the pool.

While your pool fills you can install the filter and complete the plumbing, if any. Although your home-built swim pool could easily have electric lighting, we don't recommend that you attempt any wiring around the pool. Too dangerous.

Speaking of danger, every swimming pool should be fenced to keep out and protect uninvited swimmers. Make the fence an unclimbable one.

Maintaining a healthy swimming pool is a subject in itself. Read up on it or seek the advice of local public health officials. Properly maintained, your concrete block swimming pool will give many times the hours of family pleasure that it took you to build it.

LANDSCAPING

As mentioned in the comments on Site Selection, overhanging leafy trees do add to your cleaning problems, but a stand of evergreens a fair distance from the pool will add charm, provide privacy, and cut down on unwanted breezes. Plan carefully.

Colored concrete surface cast integrally using the two-course method is combined with exposed aggregate to tone down the glare of concrete in the sunlight around pool area.

GETTING COLOR IN CONCRETE PROJECTS

Color helps set your project apart from those of ordinary gray concrete

DID YOU KNOW that there are four ways to get color into concrete as you make it? They are integral solid, integral topping, dry-shake color and colored surface aggregate. And that there are two ways to put on color after the project is built? These are staining and painting.

The most satisfactory color usually is

built right into the concrete, integrally. If when you cast your own concrete, settle on the color you want, you can do it this way. When working with masonry, buy colored units and add coloring to the mortar if desired. It beats painting.

For projects that are already built, use the staining or painting methods. Paint is

By troweling designs into the surface of a concrete slab and staining the "flags" you've created in different colors, you can make an attractive entry, walk or patio.

Pre-weighed 1-lb. bags of cement color pigment are handy if you need that amount. Otherwise, buy your coloring in bulk and weigh out batches yourself. See pp. 92-93.

Small color test samples are easy to cast in 1" pieces cut from 3" PVC plastic drain pipe. Slitting the pipe makes it a snap to remove casting from the sample cylinder.

FOUR WAYS TO COLOR
CONCRETE AS YOU MAKE IT

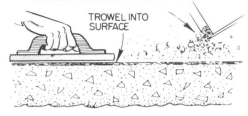

SHAKE ON COLORED POWDER

TROWEL INTO SURFACE

COLORED AGGREGATE ON SURFACE

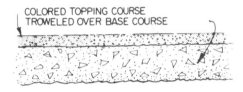

COLORED TOPPING COURSE
TROWELED OVER BASE COURSE

ENTIRE POUR IS
COLORED CONCRETE

probably the least satisfactory way of coloring concrete because of the need for repainting. If you select your paints properly, though, the work can be eased tremendously.

INTEGRAL COLORING

To color concrete integrally, do it in the mixer. Use a mineral coloring pigment. The synthetic mineral oxides produce the most lasting colors. They resist fading best and a little goes a longer way toward coloring the mix. It's tough to tell when you buy the coloring pigment, because it doesn't say what it is. All the label tells you is who makes it and what the color is. What's in the package may well be a natural oxide, which costs less than a synthetic and isn't as pure.

Nevertheless, you'll do best to buy a well-known brand of coloring pigment such as Sakrete, Davis or Pfizer.

To color concrete with the dry-shake method, dust the slab with your dry-shake coloring material just before floating. Sift it through your fingers for an even coating.

After waiting a few minutes, trowel in the color application with a lightweight metal float. Use plenty of pressure and try for an even overall color. Read text for pointers.

Oil-base concrete stains offer a greater selection of colors than dry-shake or inorganic stains do. Apply them with either brush or roller, whichever is easiest or more suitable.

Pigments are usually sold where you buy your other concrete materials. The usual package size is 1 pound. However, for big jobs you'll want to buy your coloring in 25-pound bags. It's much cheaper.

All coloring pigments for concrete come as fine powders, much like concrete itself. These are added to the mixing drum and blended along with the other ingredients in concrete. You can use them with ready mix or ready packaged mixes too.

The synthetic oxides are chromium (green), cobalt (blue), and iron (buff, biege, brown, maroon, red, black). The iron oxides are lowest in cost. Blue and green cost lots more. Black goes furthest. You can use the pigments straight as they come from the package or mix them to get in-between colors.

Don't try to judge a pigment's coloring ability in the dry state. When made into concrete, it can look quite different.

If you want clean, bright colors, always make colored concrete with white cement, instead of the usual gray portland cement.

If you want really light colors, use white sand in place of brown sand in the mix.

HOW MUCH PIGMENT?

When adding color pigments to a mix use these percentages by weight of cement:
Pastels—1½ percent
Full colors—7 percent
Never use more than 10 percent pigment by weight of cement. With black, normally half these amounts may be used.

Here's how system works: Figure how much weight of cement is in the mix. Suppose you've rented a one-third-bag mixer. This would need 31 pounds of cement per batch. The rest would be sand, stones and water. Then if you want to produce a fully saturated color you'd use 7 percent of 31 pounds, or about 2 pounds 3 ounces of synthetic oxide pigment per batch. Weigh out a number of batches each containing this amount. Use a postal or baby scale; each color batch in a sandwich bag.

To color a cubic yard of ready mix, you'd figure this way: there are 6 bags of cement (94 pounds each) or a total of 564 pounds of cement in the cubic yard of concrete. Taking 7 percent of this gives about 40 lbs. of pigment. That's a lot.

With ready packaged mixes use a maximum of 1 pound of pigment to a 60-pound bag of mix, 1½ pounds to a 90-pound bag. For well saturated colors, use 12 ounces of coloring. Use half this amount of black.

Never use anything but a concrete coloring pigment in your mixes. Paint colors, fabric and food dyes won't work.

Curing colored concrete is touchy. If you cure by covering with a polyethylene sheet, you'll get spotty colors where the sheet wrinkles. Instead, cure with sprayed-on membrane curing compound or clean, wet sand. Dirty sand will stain color.

The less finishing the better on colored projects. One, at most two, floatings followed by one troweling should be it.

Concrete coloring pigments are also good for coloring mortar. Use a maximum of 8 pounds coloring per sack of mortar cement. With ready-packaged mortar mix use a maximum of 2½ percent coloring per dry weight of mix. An 80-pound bag of mix generally gets 1½ pounds of color added. This produces rich colors. For faint colors use about one-fourth this much.

COLORED TOPPING

If you're paying more than a dollar a pound, say for green coloring (it's expensive), you begin to think there MUST be a better way. There is. Color only the top one-half to one inch of the slab. This is the two-course coloring method as shown in the drawing. It lets you build a low-cost large colored concrete project at reasonable cost for the color.

To do it, first build a structural slab of plain concrete. Strike this off ½ to 1 inch below the top of the forms. Follow immediately with a fully colored sand-mix topping course placed over the base and finished as desired. Instead of laying the topping course right away, you can place it much later, if you scratch the base course before it hardens. Later, when you are ready to place the topping—even weeks later—thoroughly clean the base. Scrub a cement-water grout into the surface of the base course to create a good bond between courses. Pour all before coat dries white.

DRY-SHAKE COLORING

Rather than have your color course pigment mixed in with the concrete, you can dust it onto the surface of a slab and float the color in. This is another economical coloring method. The troweled-in dust-on color layer is so thin—about ⅛ inch—that a little pigment goes a long way.

The colors are the same ones that are used for integral coloring. But with the dry-shake they are applied to slab's surface.

Two firms I know of that make dry-shake colors are: Master Builders, Cleveland, Ohio 44118 (*Colorcron*); and A. C. Horn Div., Dewey & Almy Chemical Co., 62 Whittemore Ave., Cambridge, Mass. (*Colorundum*).

Buy your dry-shake colors this way if

PAINTS FOR CONCRETE AND MASONRY

LOCATION	RECOMMENDED PAINT	SURFACE PREPARATION	APPLICATION	EXCEPTIONS & ALTERNATES
Exterior walls of block, brick, stucco or concrete.	Latex paint, exterior type.	Fill porous surfaces with grout and prime. Have damp before first coat.	Self-priming. Use 2 coats.	Use portland cement paint instead on very rough or moist surfaces.
Inside precast projects and walls.	Latex paint, interior type.	Same as above.	Self-priming. Use 2 coats.	Portland cement paint is good for walls, if preferred.
Outdoor precast projects.	Top quality oil-base enamel.	Have clean, oil-free and dust-free.	Prime with latex paint. Give 2 coats of enamel.	Where a glossy finish is not wanted, use latex paint for ease of application, long life.
Cinder blocks or slag blocks.	Chlorinated rubber (guards against staining).	Clean and dust-free.	Self-priming. Use 2 coats.	Portland cement paint may be used, if preferred.
Floors, slabs, patios, walks, swim pools.	Chlorinated rubber floor and deck paint.	Acid-etch new or glossy surfaces with 10% muriatic acid solution.	Self-priming. Use 2 coats.	For heaviest wear, use a catalytic coating. Portland cement paint may be used on pools.

Pigmented oil-base stains are the only ones that will cover patches and other imperfections in old concrete. They are especially formulated to be alkali-resistant.

Brush application of paint to masonry is at its best on smooth surfaces. By brushing in different directions you can get paint into the joints and rough texture of the blocks.

Rolling paint onto masonry calls for a thick-nap roller to get paint into the texture of the joints and masonry units. Latex paints can go on without any priming as shown below.

you can. If you can't buy what you want, make your own. Mix 2 parts white portland cement, 2 parts fine (mortar) sand and 1 part mineral pigment of the desired color. Portions measured by weight.

Figure on about 50 pounds of dry-shake mixture for each 100 square feet of surface to be colored. Dry-mix the ingredients before you use them. Make enough at one time to do the whole job.

USING DUST-ON COLOR

Dry-shake color can be used only on exposed surfaces of slabs where you can get at the surface to color it. They're no good for coloring surfaces cast against forms. Use the integral method for that.

Place and strike off your slab as usual. When the slab is ready for floating, sprinkle the first coat of dry-shake mixture as evenly over the surface as you can.

After waiting a few minutes for the mixture to take on water, trowel it in. Don't trowel so much that water surfaces to dilute, it will make the color spotty.

Follow the first troweling right away with a light dusting. Trowel that in too.

Run all the edges that are to be edged. as for an integrally colored slab.

STAINING CONCRETE

Even if you've heard of dry-shake coloring, bet you haven't heard of staining concrete. There are two types: inorganic chemical and oil-base pigmented stains. Chemical stains react with alkali.

Chemical stains don't produce bright

Spraying paint onto rough masonry is the easiest method, although it uses the most paint. Without trouble you can get paint into even the most intricate wall textures.

colors. The colors they do make can be enhanced by waxing with a color-matched wax. Waxes are sold by the stain manufacturer for use with its products. Directions for staining and waxing are included. Occasional rewaxing is usually needed to keep colors bright.

Inorganic chemical stains are not much good on old concrete surfaces. They won't hide patches and other imperfections. Concrete to be stained with them should be less than a year old, if possible. Colors will vary anyway. Never try to stain over a pigment-colored concrete surface.

If your dealer doesn't have the stain you want, write to Rohloff & Co., 918 North Western Ave., Los Angeles, Ca. 90029. Ask about *Kemiko* stain. It comes in eight colors: browns, beiges, greens and black.

PIGMENT STAINS

Oil-base stains, much like those used to stain wood, may be used on concrete. Those especially made for it are designed to be alkali-resistant. Cabot and Rez both produce them. You may brush or roll these stains on. Buy them at paint dealers.

Ordinary wood stains also can be used on concrete if you first knock out the alkali. Do it by applying a dilute solution of zinc sulfate (mixed at the rate of 2 pounds per gallon of water). The concrete to be stained should be clean and dry.

PAINTS FOR CONCRETE

A good many paints are made for use on concrete. You can choose from among latex, chlorinated rubber, oil-base, alkyd, portland cement and catalytic coatings (epoxy, etc.). Don't choose haphazardly, though. Have a reason for using each one. The table on page 93 shows which paint is recommended for each job.

Latex paint—especially *acyrlic* latex, which most are—beats all for general all-around use. It's no good for garage floors, however. Auto tires resting on it tend to lift it off in spots. Chlorinated rubber floor-and-deck paint has all the marbles for long wear on floors. Chlorinated rubber is a good swimming pool paint too.

The best wearing of all floor paints are the two-part catalytic coatings. Outdoors, they tend to dust off heavily (chalk), so use them indoors only. They cost quite a bit. Pick a color you can live with.

Portland cement paints are good for walls, but should not be used on floors. They are simply made of white portland cement mixed with a synthetic oxide color pigment. The color selection is thus somewhat limited. You can make your own portland cement paint by mixing white portland cement and coloring pigment about five to one: 5 cement, 1 pigment.

PAINTING

Any concrete or masonry surface to be painted should be clean, free of oil and dust-free. Always follow the specific directions on the label of the paint you use. They're the best instructions you can get.

Floor-painting goes easier if you use a long-handled extension on your roller.

Previously painted surfaces must have all old, loose paint removed before you can paint them. A good way to do this is with a wire brush or a wide scraper, or even both.

A good, quick-setting filler for repairing holes and spalls in concrete floors to be repainted is two-part catalytic auto body putty. Mix up just enough to make the repair.

Spatterdash travertine finish provides excellent footing for a pool deck. It's made by flinging a colored cement-water topping onto the slab, then troweling once lightly to finish off.

TRY THESE FINISHES AND EFFECTS

Concrete is so versatile that it takes on any creative texture you can put in it

FOR SOME USES a troweled-smooth concrete finish is best. It cleans easiest, won't scrape up your body and is hard and dense. It's the most practical basement floor finish there is. Away from the basement floor, many other concrete finishes are more suitable. A good deal prettier too.

This chapter is intended to acquaint you with some of the more unusual finishes and show you how to put them on the things you make.

There are really four ways that interesting textures and patterns get into the surface of concrete: they're tooled in, cast in, spattered on and brushed off.

BROOMING

Tooling may be done in many ways. One of the most popular and useful textures is made by brooming the concrete before it sets hard. The timing of brooming and stiffness of the broom determine the texture created. A stiff broom and early application make rougher textures. A soft broom and later application make finer textures.

Brooming may be done after a final steel-troweling when the concrete is quite hard. This makes a wispy, smooth finish. But most often brooming is done shortly after final floating with a wood or metal float. Practice brooming until it produces

Final touch in making of tooled-in flag-stones in a concrete slab is to smooth the tool marks with a wet brush. Stained different colors, the "flags" look almost real.

Portland Cement Assn.

By moving the metal float in arcs while finishing a slab, a series of swirls is made all over the surface. Swirling may be done with a wood float, metal float or a steel trowel.

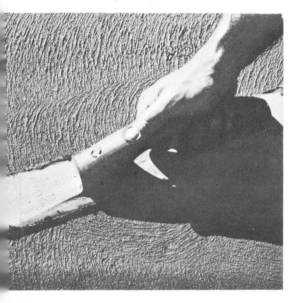

Robert Cleveland

Completed swirl finish stands out when lighted from behind. It's excellent for driveways, walks, patios where a nonslip footing is needed in bad weather for cars and people.

the effect you want. Unsuccessful tries can be troweled over to smooth the roughened concrete.

TEXTURED FORMS

Casting concrete on materials that are textured, such as weathered plywood, makes similarly textured concrete. In general, anything you cast concrete against will impart its surface texture to the concrete. Concrete cast on glass is glass-smooth. Concrete cast against wood, takes on a graining just like the wood. Concrete that's cast on a car floor mat looks like the floor mat. Rubber matting with intriguing textures is available from Boston Woven Hose Div., American Biltrite Rubber Co., Box 1071, Boston, Mass. 02103. Plastic form liners are supplied through Monsanto Chem. Co., Plastics, Springfield, Mass.

SPATTERED-ON TEXTURE

What's called a travertine finish resembles travertine marble in color and texture. It's made in two courses. The first course is finished level with the tops of the forms and scratched to roughen it for a good bond between courses. A cutoff piece of hardware cloth makes a good scratcher.

Portland Cement Assn.

Brooming done early in the setting process like this makes a deeply textured finish. Done later or with a softer broom, the texture is milder. Broom in any direction.

Portland Cement Assn.

Moon-like texture is made by casting concrete stepping stones face down on polyethylene sheeting laid over pebbles. The result is a surface that's rough but glossy.

The second cement-color-water coat is applied at whipped cream consistency after the first has cured several days. A wallpaper brush is used to fling it down onto the grout-scrubbed surface. A light steel-troweling levels out the peaks and fills in some of the valleys.

A travertine finish, while most attractive, is not recommended for use in exposed locations in freezing climates. Water tends to collect in the depressions, freeze and pop off some of your topping along with it.

EXPOSED AGGREGATE

A beautiful colorfully textured finish you can use in any climate is exposed aggregate. We've mentioned it in other chapters for use on your around-the-house projects. Now here's how to make it.

Casting concrete against wood-board forms reproduces the surface of the boards on the finished wall. Alternating board directions at V-strips simulates a checkerboard effect.

Some of the most intriguing looking shapes are those cast in vacuum-formed plastic molds. To avoid surface air bubbles, the molds should be vibrated after filling.

Portland Cement Assn.

Concrete cast against rubber matting or stair treads has a terrific texture for use in unusual stepping stones. People will ask where you got them. Oil mat for easy release.

Portland Cement Assn.

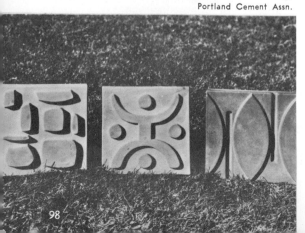

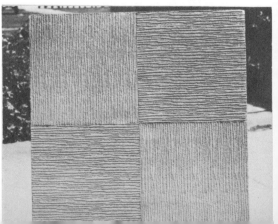

The process is a little different for a slab than for a surface cast against a form. First the slab.

Cast your slab in the usual way, but with a colored concrete. Ordinary gray concrete seldom looks like much in exposed-ag projects.

For the tightest ag patterns cast the slab in two courses. The top one should be sand, cement and coloring, but no water. The bottom one can be plain gray concrete. Make the top course at least twice as thick as the diameter of your largest aggregate for exposing. This is to allow depth for pressing the ag into the topping.

Float the slab once and leave it. When the surface is "thumbprint" hard, you're ready to proceed. This is when you can press your thumb in and make an impression without getting it buried beneath the surface.

SPREAD PRETTY AGS

Attractive, colorful stones can be purchased in bags or picked out of aggregate piles. They should be approximately all one size or two sizes at most. All should be hard stones suitable for use in concrete.

The best place to look for attractive ags, if not along local streambeds, is in the aggregate pile at a materials supplier. Ask if you may pick out pretty stones from the owner's piles and weigh them up. Bring the kids and make it a family project.

You can buy select washed stones in 100-pound sacks. Look for them at a large concrete products supply yard or at a terrazzo supply house. You can get all one color, or perhaps choose a second color to distribute as a color accent. The larger the stones the rougher the texture your finished project will have. Yet there's nothing to prevent you from using 4-inch head knockers if that's the effect you want.

Dampen the stones and spread them in a one-deep layer over the slab surface. Arrange some stones by hand, if necessary to get a good spread. Then tamp the stones into the still-soft concrete with a float. If the mix is too stiff for floating-in the ags, try tamping them in with a 2x4 held on its side. If it's too stiff for this, you waited too long. Try tamping with a brick to get around the problem. If you spread and float too soon, the aggregate will sink below the surface of the slab. It usually

Portland Cement Assn.

One step in making an exposed-ag finish is to spread select aggregates evenly on the surface. Then float in. Later the mortar covering is brushed and flushed off surface.

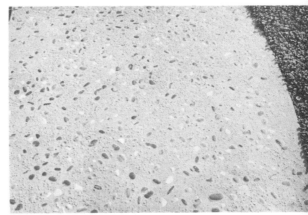

Robert Cleveland

The resulting surface contains colorful stones with their tops showing. Note the wide spacing between stones. This is because a sand-mix topping was not used.

When sand-mix topping is used to make exposed-ag concrete, the select aggregate can be very close together. There are no other stones in mix to hold them apart.

99

To make a rock salt finish, spread salt crystals over the freshly floated slab and trowel them in flush with the surface. This should be the last step before curing the concrete.

Depending on how thickly you spread the rock salt, the surface will be covered with pock marks when the salt dissolves out. The slip-resistant surface isn't freeze-durable.

takes about one hour after strikeoff, but watch out on a hot, dry day. Setting comes fast.

Float or tamp only until the aggregate is embedded flush with the surface of the slab. Then stop. If it goes below the surface you'll lose it.

BRUSH AND FLUSH

Now wait again, this time until the concrete can be brushed and hosed without dislodging any stones. How long this takes

will depend on temperature, and how stiff the original mix was. Figure about another half hour or so. When conditions are right, you'll be able to hose the slab and broom it with a pushbroom gently and remove the covering of mortar from the stones. If you start too soon, you'll wash out stones. If you wait too long, you'll have to use a wire brush in place of a broom to make any headway. The concrete will be stiff enough for you to get onto it at this point. Walk as gently as you can to avoid dislodging stones.

Concrete retarder is used to get an exposed-ag finish on cast concrete objects. Brush it evenly over the form insides and let dry. A milk carton serves as an easy-make form.

After the form has been stripped, the outer mortar covering is still soft because of the retarder. Simply brush it away exposing the select aggregate down to the desired depth.

To cast exposed aggregate tiles face down use the sand-embedment method. Set the form on a thin layer of sand. Sprinkle the ag in sand and press. Cast concrete over it.

Robert Cleveland

Aggregate for exposing needn't be small. Here cobblestone-size ag has been set in Marine drill order, surrounded by concrete and later exposed by brushing and flushing.

Expose aggregate until you get the appearance and texture you're after. Don't expose more than about the top half of the stones. The top third is better. The stones must be embedded enough that they'll stay in under traffic.

Give the slab a final rinse. To really get it bright follow a day or two later with acid-etching. Use 1 part muriatic (hydrochloric) acid in five parts water. Wear rubber gloves and goggles. Flush off the acid with water as soon as it stops foaming.

Don't try to take on more than about

50 square feet of exposed aggregate slab by yourself. On a hot day in the sun, it can quickly get ahead of you.

CASTING EXPOSED-AG

Since the surfaces of cast objects aren't usually exposed for brushing until you remove the forms, use a different system to make them.

If the project is small, make the mix using cement, sand, coloring pigment and the attractive aggregate. When the mortar

Incised-in tool marks make good looking tiles for walks. The designs can be made with tin cans, cookie cutters, trowels or almost any other handy tools you want to use.

Hardened tile picks up stones leaving them exposed where they were buried beneath the sand. This shape, while pretty, is a poor one for tiles. The arms tend to break off.

Impressed designs in slab surfaces are becoming popular. These were done with aluminum "brick" molds. You can get a similar effect using a wooden imprinter.

Robert Cleveland

Boulders were set between the forms for a sidewalk, then exposed-aggregate concrete cast in them. During exposing, the boulders were exposed too, leaving a variety effect.

Press-in patterns made with various-sized tin cans are easy to get and decorative too. Have the slab ready to cure so you won't have to trowel or finish over your designs.

Portland Cement Assn.

on the surface is brushed away, the ag will show. If the project is large and would waste costly pretty aggregate, cover the surfaces of the form with the "pretty" mix first. Then follow with a plain mix, being careful not to poke the plain mix through the pretty mix at any point with your handling.

Another method is to spread the attractive ags against the form and pour colored sand-mix concrete over them. This works well only when casting things like paver tiles upside down. Vertical surfaces can't get exposed-ag in this way unless you figure a way to glue them to the form temporarily. We've experimented with flour paste, white glue and contact cement without great results. Got any other ideas? Professionals use a two-part epoxy substance. Use your imagination.

USE OF RETARDER

To keep the mortar next to the forms soft until you can remove the forms and expose by brushing, use a paint-on substance called a *retarder*. Brush it onto the form and let it dry. Then any concrete placed in that form will set up on the interior, but stay soft on the outside. You can strip the form after a day or two and the mortar covering will brush away easily.

It takes a little experimenting to get the timing right. If you strip and brush too soon, your aggregate pretties will slough away soon as they lose the support of the form. If you wait too long, the mortar covering will set hard, retarder or not. You'll have a tough time exposing any aggregate.

Another problem: the set-retarding effect of the retarder depends on how thickly you put it on. Unless you are careful during application, you can have parts of your project ready for exposure, portions past the exposure time and still other parts not ready for exposing.

The best procedure is to experiment. If you go at it with learning rather than perfection in mind, you can't help but be successful—at least in learning.

There are a number of brands of retarder for form application but none are readily available at do-it-yourself outlets. To get it write to Burke Concrete Accessories, Inc., 3870 Houston Ave., San Diego, Cal. 92110. Ask about *Agreveal-F*. They will be glad to help you.

TIRE THAT MAKES A TETHERBALL GAME

Here is a boy-sized game that is sturdy yet easily portable, and easy to make

Concrete-filled-tire base makes tetherball game heavy enough that it won't tip over. It can be set up in a play yard, driveway or sidewalk to save lawn. Tire is protection.

"**H**EY YOU GUYS! I've got a great idea! Let's play tetherball."

"Won't your Mom get mad like last time if the baby's asleep and we make a lot of noise?"

"Well then we can roll it over to your house."

"Whaddaya mean?"

"Dad made us a neat new tetherball game out of an old tire and some concrete. And we can roll it anywhere we want and set it up."

"No kidding! Let's see it."

"It's in the back yard. Dad put a piece of plywood across the big hole in one of his old car tires so it would hold concrete. Then he poured it full. Before it got hard, he shoved a piece of pipe in and got it sticking straight up.

"Now we can roll the tire part anywhere we want and stick our tetherball pole inside the pipe and play. Dad says that way we don't wear out the grass in one spot like we used to."

"Hey, neat. Can we take it over to my house? Maybe my dad'll see it and make one for us."

To keep the base from being too heavy, the old tire was stuffed half-full of crumpled-up newspapers before casting. A plywood disc sealed off leaks around the bottom bead.

The 1¼" receptor pipe is cast into the base to hold the 1" tetherball pole snugly. As soon as the pole has been inserted in base, game is ready to go—almost anywhere.

105

SAND-CASTING CONCRETE

Damp sand makes a pliable casting mold

for many of your artistic impressions

S AND-CASTING IS AN ART FORM in itself. All we can do here is introduce you to its possibilities. Then let the creative artist in you take over. Give it free rein and you may be amazed at the things you come up with as house decorations.

To make a sand-casting, hills and depressions are made in a bed of sand and concrete is poured onto the bed. The concrete picks up every detail you put into the sand mold but in reverse. Valleys made in the sand become mountains in the casting, and vice-versa.

Those of us who haven't much artistic talent can still enjoy casting in sand molds by copying what we see. Sort of sand-cast-by-numbers. We can also push things into the sand and make castings of their impressions the way the FBI does when they find a kidnapper's tire tracks. In fact you could make a kind of "Crossroads of America" wall plaque by rolling different-sized tires across some sand and laying a frame around it to cast in. Impressions of a car snow tire, bicycle tire, lawnmower wheel, motorcycle knobby tread and wagon wheel all could be included.

Sand-casting is fun and an outlet for talent. This sand-casting by San Diego artist Charles Faust shows some possibilities. Plaster cast flower head and pebbles were embedded.

Dish-shaped garden bowls can be cast in a sandpile. A plywood template is rotated around a center pivot to make the shaped depression in which exposed-ag bowl is cast.

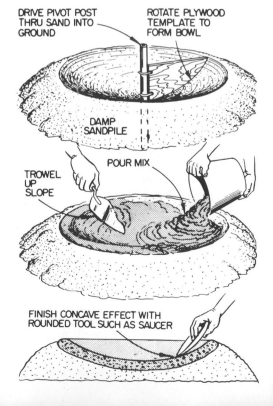

DRIVE PIVOT POST THRU SAND INTO GROUND

ROTATE PLYWOOD TEMPLATE TO FORM BOWL

DAMP SANDPILE

POUR MIX

TROWEL UP SLOPE

FINISH CONCAVE EFFECT WITH ROUNDED TOOL SUCH AS SAUCER

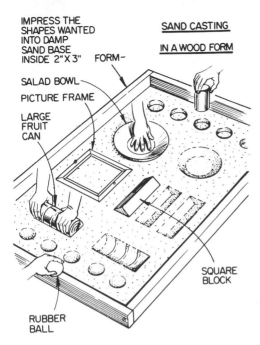

IMPRESS THE SHAPES WANTED INTO DAMP SAND BASE INSIDE 2" X 3" FORM—

SAND CASTING IN A WOOD FORM

SALAD BOWL

PICTURE FRAME

LARGE FRUIT CAN

SQUARE BLOCK

RUBBER BALL

A set of different-looking concrete parking bumpers is made by sand-casting. Form the depression with a template. Spread the exposed-ag stones, then pour the concrete.

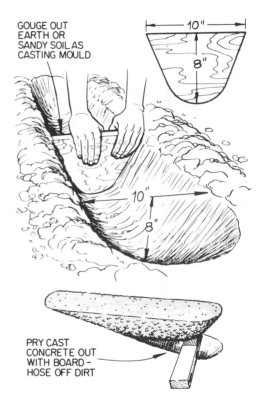

GOUGE OUT EARTH OR SANDY SOIL AS CASTING MOULD

10"

8"

10"

8"

PRY CAST CONCRETE OUT WITH BOARD — HOSE OFF DIRT

"Tools" you can use to make sand impressions are unlimited. Try using shop tools, kitchen utensils, cookie cutters, blocks of wood, your hands, elbows, feet, a twig, a can bottom, a bottle cap.

The texture your sand-casting will have depends on the kind of sand you cast it in—fine or coarse—and what you do to the surface of the sand before casting. A rough texture is made by flinging water onto the sand surface after making the impressions but before pouring the concrete. Pushing a piece of hardware cloth, fine or coarse mesh, into the sand will give the casting a checkered look. Sprinkling dry sand on top of the mold will impart a hilly but smoother texture.

You can put things onto the sand that will be embedded in the casting and show as a part of it. Stones, tiles, wood and such may be used to add different colors and textures when embedded in the casting.

There are three ways you can make your sand molds for sandcasting: Use the sandpile as its own mold. No framing is needed. Or you can level a spot on the sandpile and build a 1x1-inch wood frame and whatever impressions you make in the sand will appear on the casting. The casting will have the shape of the frame.

The third way is to build the frame extra deep, say out of 1x2's on edge nailed to a plywood bottom. Fill the frame half full of sand and level it off by screening. Cast

in the sand bedding. Strike off the concrete across the top of the frame. Using this method you don't need a sandpile. Just enough sand to half-fill the form.

Mixes for sand-castings can be anything you want. You can even use plaster, as many artists do. Any of the concrete mixes listed in the next chapter will work fine. Recommended for making artistic wall plaques are the light-aggregate mixes.

When pouring your sand-casting mix, do it gently. If you don't, the concrete will make its own impressions in the sand.

Terrazzo mix contains one or two colors of marble chips that, when exposed by wire-brushing, add an accent color to the concrete around them. The time of brushing is critical.

DECORATIVE CONCRETE MIXES

Use one when you want something a bit different from an ordinary concrete mix

THE TWO TYPES of ready-packaged concrete mixes will handle most projects you might build. However, if you want something different for a decorative project, you can design a mix of your own to suit the job. We'll help you get started.

First of all, here's a list of some projects you might make with a decorative mix: ashtray, bench, bike rack, birdbath, bookends, bowls, chairs, coffee table, end table, flagstones, flowerpot, fountain, lamp base, lantern, lawn edging, mailbox post, paperweight, pedestal, picnic table, planter

plaque, pot, statue, stepping stones, tile, urn, vase.

Here are the most useful decorative mixes. The table tells how to proportion the materials.

WHITE CONCRETE

Normal concrete made with regular portland cement is a dull, lifeless gray. This is true whether or not you mix it yourself. But you can make white concrete by using white portland cement in place of regular cement, if you have good reason.

White cement costs about twice what regular portland cement does. No brand of ready packaged mix is made with it, so you'll have to mix your own. You can probably buy white cement from a terrazzo supply house if your building materials dealer can't get it for you.

If you want a really white end-product, use white silica sand in place of regular sand in your mix. Brown sand gives a faint beige cast to your project.

TERRAZZO MIX

The great-looking terrazzo floors you see in many public buildings are made with colored concrete that contains one or two colors of marble chips. After the concrete sets, the surface is ground smooth to expose the marble chips and make the floor look glamorous. By brushing the still-soft surface with a wire brush or by disc-sanding it after overnight hardening, you can produce a similar effect without grinding. I call it *rustic terrazzo*. The surface looks like terrazzo but is a little rougher.

In making up the terrazzo mix, you can use plain cement or white cement. Or you can add the marble chips and coloring to ready-packaged sand mix concrete if you prefer. In this case the chips will be farther apart at the surface because the sand grains are between them.

Remove the form within 8 to 10 hours, being careful because the terrazzo concrete is still delicate. Then expose the marble chips at the surface with a wire brush disc sander.

The terrazzo portion of your project need be only skin deep. Place this mix against all portions of the form that make exposed surfaces. Then fill in the rest of the project immediately with regular concrete. It's cheaper than using your deluxe mix to make solid terrazzo. It'll look the

Projects cast of no-fines concrete have a pronounced texture that depends on the size of the coarse aggregate. Keep the aggregate fairly small to avoid weakening the cured concrete.

Pock-marked surface is given by incorporating rock salt in with the mix ingredients. Add the salt after the mix has blended in the mixer to keep it from dissolving away.

same on the outside anyway. The two mixes will bond together as strong as one if you mix and pour the second one immediately after the first.

Using a stiff mix helps you to strip the forms sooner and get at marble-exposing soon as possible. When casting tiles or other flat objects, often the forms can be stripped right away. You'll get a little slumping at the edges, but this needn't hurt the looks.

LIGHT AGGREGATE MIXES

Concrete for a decorative project that doesn't need much structural strength can be made with lightweight aggregates. These include vermiculite—the lightest— and such natural or manufactured light ags as scoria, pumice, Perlite, Haydite, porous slag and crushed lava. Vermiculite is the same stuff that's sold in bags for gardening and insulation.

The lightweight materials may be used as both fine and coarse aggregates. Really light ones such as vermiculite produce a concrete so soft that you can saw it or sculpture it easily with hand tools. Such a mix, however, hasn't any resistance to

DECORATIVE CONCRETE MIXES

MIX	CEMENT	SAND	COARSE AG	WATER	PROPERTIES
White concrete	1	2½ (silica)	3 (small stones)	½*	Hard, structural white concrete Stones may be omitted for sand mix.
Terrazzo	1 (white)	0	1 (marble chips)	½*	When brushed, looks like terrazzo
Vermiculite aggregate (lightest)	1	0	4 to 6 (vermiculite)	1½**	Makes lightweight concrete that can be carved. Poor weathering
Vermiculite aggregate (light)	1	2	3 (vermiculite)	1½**	Lightweight, poor weathering but smoother than above. More cement gives added detail
Light-ag manufactured or natural	1	2	3 (pumice, etc.)	½*	Makes good-weathering concrete of fair strength but lighter than normal
No-fines	1	0	9 (⅜" dia.)	½ or just enough to cover ag	Lightweight concrete covered with air spaces
Pocked	1	2½	1½ (rock salt)	½*	Mix quickly and place before salt dissolves. Makes pock-marked surface

*Use 1 part water for a "pourable" mix
**Use 2½ parts water for a "pourable" mix

freezing and thawing. Use it for projects that are protected. For outdoor projects use one of the harder light aggregates.

Light ags are thirsty for water, and so you'll need to use more water than in a regular mix.

Vermiculite produces lumpy-surfaced castings that won't pick up much form detail. While this is fine for many uses, if you want to record fine detail, add some sand to the mix. It smooths out the texture and picks up detail.

NO-FINES MIX

By leaving the sand out of a normal concrete mix you make concrete that looks like a mass of stones glued together by cement paste. That's what it is too. It's called no-fines concrete. To make it successfully you'll have to experiment with the proportion of water. Too much water makes such a thin cement paste that it won't stick the stones together. Too little water fails to coat every stone. Add water a little at a time in the mixer until every stone is just coated with paste. Then pour.

POCKED CONCRETE

Made by adding coarse rock salt to the mix, pocked concrete contains small voids all over its surface where the chunks of salt melt away. Add the salt to the mix as shown in the table. After curing, wash away any surface salt remaining.

This same effect can be had by spreading rock salt sparingly in the form before pouring concrete over it.

If you want to experiment with wild surface textures, sprinkle some baking soda in the form before you pour. It reacts with the concrete, producing void-creating gas bubbles. You take your chances on where they occur and how many there are.

Void concrete, whether made with rock salt or by some other means isn't recommended for outdoor exposure.

Mixes in the table are intended as guides to start you off. You may vary them as you wish. Portions are by volume. If all measurements are made in the same container, it doesn't matter whether the container is a thimble or a pail. A "pourable" mix is needed when filling intricate forms. Use additional water to make a "pourable" mix, as shown in the table.

MONEY-SAVING TIPS

Most people think that masonry work is expensive, and it is—but you can beat the system; here's how to do it

Robert Cleveland

If you have a chain saw and access to large dead trees, you can build an estate-sized patio for very little by using log rounds as paving. Fill spaces with concrete and it's done.

PROJECTS BUILT of concrete and masonry are usually considered worth more than the same ones of wood or some other material. Part of this is because of their long life and low maintenance. Also, labor costs for building masonry tend to run high. But when you do your own construction, you save all labor costs. And you can save more on materials.

HOW TO SAVE

Here are a number of ways you can cut down on material costs of your concrete-masonry projects without sacrificing appearance:

Use cheaper materials where they won't show. For instance, in laying up a garden wall, you can use fancy face brick on the front, switching to plain concrete block at the back. Never use costly units where they're hidden below ground. Instead lay these portions with low-cost concrete block or brick.

You can save money by using ready-packaged concrete mixes for small and medium-sized projects, switching to ready-mix or mix-it-yourself construction for large projects. If you're in doubt, figure the cost both ways and see which is most economical. Generally anything less than one cubic yard is an excellent candidate for a ready-packaged mix.

Ready-packaged mortar mixes are good choices for all but the largest masonry projects. They're much cheaper than finding and hauling in mix-it-yourself materials.

USED BRICKS

You can save dollars on masonry units by going along with what's available in your area. For instance, in some places used bricks are selling at fantastic prices, even more than for new bricks. Yet in other places with lots of urban renewal going on and thus a large supply of used bricks, you can get used bricks quite reasonably. If prices of real ones are out of sight, it may pay you to buy man-made "used" bricks. They are made to look as though they'd once been laid up in a wall even though they haven't.

Learn what kinds of concrete masonry units the block producers near you turn out. Design your projects with these in mind. Having unusual units hauled in adds to the cost of your project.

One way of saving money has been stressed strongly. Don't use reinforcing steel where it is not needed. Most non-engineered uses of steel reinforcement merely add to the cost without doing much to help the strength or durability of a concrete project.

NO WEEKEND READY-MIX

A great way to save on the cost of ready-mixed concrete, if you can swing it, is to schedule your jobs for week-days. Most ready-mix producers charge extra for Saturday deliveries because it's an extra-pay day for drivers. Saturday is a busy ready-mix day anyway.

Plan all dimensions of masonry projects

Instead of building a continuous walk, save by laying cast concrete stepping stones where needed for a walk. These units were cast against pegboard for a textured surface.

Scrape the mortar from any form lumber as it is removed. A masonry trowel makes a good scraper. If boards are not clean, your saw will dull when cutting through them.

Robert Cleveland

A low-cost stepping stone walk needn't be plain. Here the edging was made with 1"x4" wood strips staked into the ground. Precast steps were set in gravel between the edging.

Broken bricks and colored concrete blocks make a low-cost exposed aggregate material. Smash them into the sizes you want with mason's hammer and place in form.

Plan to re-use concrete form lumber. As soon as you finish a concrete project, scrape the mortar covering from the form tops with the edge of your finishing trowel.

To make exposed-ag stepping stones with the crushed topping, simply pour concrete over it in the forms and screed it off. Top of form will be bottom of stones, chips, etc.

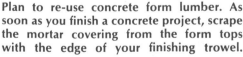

When the stepping stones have hardened for about a week, tap them out of the form with a rubber mallet. The crushed brick material is neatly embedded in the surface.

to work out in even numbers of units with the least possible cutting. For example, when using 8x8x16-inch concrete blocks, dimensions in an even number of feet will always work out in full- and half-blocks. The same is true of dimensions in an even number of feet plus 8 inches. Dimensions in an odd number of feet will not work out. But an odd number of feet plus 4 inches will always work out in full- and half-blocks. These rules hold true in both the horizontal and vertical directions.

Whatever you do to save money on concrete-masonry work, don't do it at the expense of a good job. Don't skimp on concrete or mortar quality or masonry unit quality to save a few bucks. You'll pay for it later. Your best bet is to buy known brands from reputable local dealers. That's good advice anytime.